How Economics Became a Mathematical Science

Science and Cultural Theory

A Series Edited by Barbara Herrnstein Smith

and E. Roy Weintraub

How Economics Became a Mathematical Science

E. Roy Weintraub

Duke University Press   Durham and London 2002

© 2002 Duke University Press

All rights reserved

Printed in the United States of America on acid-free paper ∞

Typeset in Stone Serif by Keystone Typesetting, Inc.

Library of Congress Cataloging-in-Publication Data appear

on the last printed page of this book.

For my brother, Neil
(November 1, 1951–August 16, 2000)

Contents

Mathematics is the queen of the sciences.
—*Karl Friedrich Gauss*

It seems to me that no one science can so well
serve to co-ordinate and, as it were, bind
together all of the sciences as the queen of
them all, mathematics.—*E. W. Davis*

The body of knowledge includes statements
that are the answers to questions related to the
subject matter of the given discipline. The
images of knowledge, on the other hand, include claims which express knowledge about
the discipline qua discipline. . . . Thus images
of knowledge cover both cognitive and normative views of scientists concerning their own
discipline.—*Leo Corry*

Acknowledgments

Several chapters, or portions of chapters, have earlier appeared in a somewhat different form in other publications. I gratefully acknowledge the various publishers' permissions to reprint, with some significant revisions, those materials here. Portions of chapter 1 from "Measurement, and Changing Images of Mathematical Knowledge," in Mary Morgan and Judy L. Klein, editors, *Measurement in Economics* (Supplement to Volume 32, *History of Political Economy*) Durham: Duke University Press, 2001; chapter 2: "From Rigor to Axiomatics: The Marginalization of Griffith C. Evans," in Mary S. Morgan and Malcolm Rutherford, editors, *On The Transformation of American Economics, From Interwar Pluralism to Postwar Neoclassicism,* (Supplement to Volume 29, *History of Political Economy*) Durham: Duke University Press, 1998, pp. 227–59; portions of chapter 3 from: "Axiomatishes Mißverständnis," *The Economic Journal,* 108, November 1998, pp. 1837–47; portions of chapter 4: "The Pure and the Applied: Bourbakism Comes to Mathematical Economics," with Philip Mirowski, *Science in Context,* 7:2, 245–72, 1994; chapter 5: "Negotiating at the Boundary: Patinkin v. Phipps," with Ted Gayer, *History of Political Economy,* 32, 3, Fall 2000, 441–71; chapter 6: "Equilibrium Proofmaking," with Ted Gayer, *Journal of the History of Economic Thought,* 2001; portions of chapter 9: "How Should We Write the History of Twentieth Century Economics?," *Oxford Review of Economic Policy,* 15, 4, Winter 1999, 139–52.

I began writing this book in 1992 while visiting the University of Venice, Ca' Foscari. Very early drafts of three of the chapters were improved by conversations that fall with Lionello Punzo and Giorgio Israel.

The project went on a two-year "hiatus" while I served as Acting Dean of the Faculty of Arts and Sciences at Duke University. Beginning anew in 1996, I received comments on early versions of several chapters in a number of seminars, conferences, and workshops. For this I thank Ivor Grattan-Guinness, Andy Warwick, Marcel Baumans, Mark Blaug, Harro Maas, Margaret Schabas, Alan Hynes, Robert Leonard, Grant Fleming, John Lodewijks, Michael White, Peter Boettke, Cristina Marcuzzo, Bruna Ingrao, Donald Katzner, and Randall Bausor.

In the e-connected world of scholarship, virtual colleagues Phil Mirowski, Esther-Mirjam Sent, Wade Hands, and Warren Samuels were willing to make helpful comments at a distance.

My Duke colleagues Neil DeMarchi and Craufurd Goodwin were most patient as I finally worked out the arguments of the volume in classroom and HOPE workshops.

Ted Gayer and I collaborated on two of the pieces that are here presented as chapters; I could not have completed this project without his enthusiastic support.

At the late stages of this project, I received comments on specific chapters from Simon Cook and Leo Corry, while Mary Morgan and Roger Backhouse provided extensive comments on the entire manuscript. I am immensely grateful to them for giving me their time and assistance.

The book took final form during a sabbatical leave of absence from Duke in 1999–2000, with time split between the University of Rome "Tor Vergata," and the University of Nice-Sophia Antipolis (IDEFI). I thank those two institutions for their hospitality.

My Duke Press co-editor Barbara Herrnstein Smith has been a source of support in this long project, and an intellectual beacon shining light on my path. Reynolds Smith, for Duke University Press, reconfirmed the value of a good editor. In creating the Economists' Papers Project at the Duke University Special Collections Library, Director Robert L. Byrd gathered and made accessible the archival records on which this volume is mostly based; our discipline is in his debt. Joan Norris Shipman transcribed the rough tapes with which I begin each part of any writing process and it was thus her competence in making sense of the unintelligible that forced me to stop avoiding the work of writing this book.

The material in chapter 7, based on my father's letters, was difficult to organize and make coherent. I owe much to my late brother Neil for his emotional support for my constructing that account, and am profoundly grateful that he had a chance to read it shortly before he died.

Finally my wife, Nell Maxine Soloway, never flagged in her understanding of this book's demands, nor did she waver in her support for its author. "Thank you" barely acknowledges my debt to her.

Prologue

History of mathematics is history of knowledge. But this history is a social process and "knowledge" has to be taken in the widest sense of the term. . . . An open-minded historiography will help to widen our understanding of knowledge and mathematics, and of the social processes of their historical development.
—H. Mehrtens, "Social History of Mathematics"

Neither economists nor historians have produced a serious and detailed analysis of the changing views of mathematicians and economists over the last century on the use, nature, and meaning of mathematical economics. Instead, studies of the connection of mathematics and economics have considered one mathematical economist or another or one problem or another (e.g., general equilibrium analysis, international trade theory). Alternatively, there have been studies of the success or failure, in principle, of mathematics in economics as in "Resolved: There is too much math in economics" or "Loose thinking results from informal thinking." What we do not have is a systematic investigation of the economics profession's engagement, or putative engagement, with the ideas of the community of mathematicians in the twentieth century.

I employ a distinction developed by the historian of mathematics Leo Corry, which he acknowledges having adapted from the work of the historian of science Yehuda Elkana. In a series of articles (Corry 1989, 1992a, 1992b), and more fully in a book (Corry 1996), Corry explored the difference between what he called the corpus of mathematical knowledge, and the image of that mathematical knowledge:

We may distinguish, broadly speaking, two sorts of questions concerning every scientific discipline. The first sort are questions about the subject matter of the discipline. The second sort are questions about the discipline qua discipline, or second-order questions. It is the aim of the discipline to answer the questions of the first sort, but usually not to answer questions of the second sort. These second-order questions concern the methodology, philosophy, history, or sociology of the discipline and are usually addressed by an ancillary discipline. (1989, 411)

The first sort of question concerns the discipline's knowledge, while the second concerns the image of knowledge. Corry's argument is that to speak of change in mathematics is to speak not only of change in mathematical knowledge, in the sense of new theorems proved, new definitions created, and new mathematical objects described. But change in mathematics also involves changes in the image of mathematics, in changed standards for accepting proofs, changed ideas about rigor, and changed ideas about the nature of the mathematical enterprise. For Corry, "it is precisely the task of the historian of mathematics to characterize the images of knowledge of a given period and to explain their interaction with the body of knowledge—and thus to explain the development of mathematics" (Corry 1989, 418).

My own perspective is shaped by this argument of Corry's. If economics is intertwined with mathematics in the twentieth century, in order to understand the history of economics we need to understand the history of mathematics. However, the history of mathematics involves the history of both changes in mathematical knowledge and changes in the images of that mathematical knowledge. Thus in what follows I will refract economics through the prism of changing images of mathematics. My reconstruction of the development of a mathematical economics will not much involve the excavation of the mathematical knowledge—theorems, definitions, concepts—that has been imported into economics. Rather, each chapter will explore, more or less directly, how economics has been shaped by economists' ideas about the nature and purpose and function and meaning of mathematics. My argument, which will be reprised often, is that one can tell a coherent story about the development of those ideas in the community of mathematical economists by attending to the evolving image of mathematics held by the community of mathematicians.

I submit that an internalist history of economics, a history of the form "Economist A's contributions begat Economist B's contributions begat, etc.," cannot make good sense of the reconstruction of economics as an applied mathematical science in the twentieth century. Historical confusion results, in part, from the failure of economists themselves to understand the changing image of mathematics. It will become apparent too that historians of economics, even as they have considered the views of one or another economist about mathematical economics, have failed to recognize, or question, the assumption that the nature of mathematics is fixed and monolithic even as the stock of theorems increases each year. Although the proposition has been denied in recent years by serious historians of mathematics, historians of economics (and most economists) continue to believe that mathematics is somehow there, and will always be there in but one shape and form even as over time the corpus of true theorems expands.

Moreover, the usual historiographic perspective is nearly always focused on particular mathematicians:

> Studies, both ancient and modern, in the history of mathematics have often tended to focus exclusively on the contributions of various individuals to a vast body of mathematical knowledge, while giving little or no consideration to the cultural context in which the mathematical developments took place. Whereas such works clearly play an indispensable role in documenting the growth of mathematical ideas, they also serve to reinforce or perpetuate a widespread myth that the history of mathematics involves nothing more than recording the "discovery" of completely disembodied ideas. (Parshall and Rowe 1994, 295)

Although I will construct a number of my own stories about change by pairing particular mathematicians with particular economists, my pairings point to ways of reading the economist against the history of mathematics itself. Moreover, my "people" will not always be the "great" through whom the standard narrative travels. Indeed some individuals' lack of high status will reveal with unusual clarity the contentiousness of the underlying issues.

Even as economists remade economics in the middle third of the twentieth century by employing mathematical ideas of optimization theory, game theory and programming, dynamical theory, and probability theory, they also struggled to assimilate new ideas about axiomatics, formalism, and rigor, as professional mathematicians reconstructed the meanings of those terms. In the history of economics literature, this remaking of economics is a story of the continuous movement of ideas from mathematics to economics. In what follows I will reconstruct a different history.

There are many ways to tell the story of the mathematization of economic theory, and connected to each of those ways is a reason, perhaps more than one reason, to tell the story in that way. The stories we tell are not natural. The evidence we bring to bear is not obvious, and the narratives we construct are not imposed by that evidence but instead are constructed along with the evidence we select to bring out in our accounts. However, unnaturalness is not usually the reason why any particular account fails to convince any particular reader. Instead, it is the failure to make the unfamiliar past coherent that leads to dissatisfaction with one or another chronicle of a time past.

My response to this historiographic challenge shapes my story, as does my desire to construct a useful narrative for economists, historians of science and mathematics, science studies scholars, historians of economics, and all their students. It is in that spirit that the following chapters might best be read. They provide a perspective on the twentieth century's economics that allows me to ask some questions and to tell some stories that, as they engage your interest, may change your beliefs, and enable you to tell your own new and interesting stories.

Many years ago, Bloomsbury's Lytton Strachey reconfigured historical writing with his *Eminent Victorians*, a book that was in fact a set of four essays on individual public figures from the Victorian era. My own chapters are mostly self-contained, but they jointly tell a layered story of a period. Although they are not such an eloquent Strachey-like debunking of historical verities, I believe that their combined weight destabilizes the traditional history.

Several of the chapters have appeared as journal articles even as they were originally conceived as parts of a whole. Those pieces, earlier revised for separate publication, have now again been revised, most of them substantially, to stand in a more transparent relationship one with another to construct a story of the interconnection of economics and mathematics over the first two thirds of the twentieth century. The reader will see that their serial concerns follow roughly a chronological order, with each presenting a temporal context for succeeding chapters.

Historiographically I have been developing my ideas over time. From a fairly traditional Lakatosian perspective favoring rational reconstructions of historical material, which guided my thinking through the 1980s, I have in recent years been (not so much guided as) cognizant of the value of writings in science and technology studies. This more open historiographic perspective has allowed me to pursue a variety of issues about the development of mathematical economics closed off from, or inaccessible to, more traditional investigative logics.

The material that follows continues this journey. And the story of course, as is necessary, reflects both personal history and the social context of the intellectual history. My father was an economist, one who trained in the 1930s, and did his major work in the postwar period, passing away in 1983. His mathematician brother represented (as I try to suggest in chapter 7) a set of intellectual connections that shaped some of my father's thinking. As a six-year-old, I received a copy of my father's *Price Theory* with the hand-written inscription, "To Roy, who will someday write a better book." My father, who had read John Stuart Mill's *Autobiography* with care, who had been enchanted by the materials in Harrod's biography of Keynes concerning the intellectual relationship between John Neville Keynes and son Maynard, and who was later to spend a great deal of time talking about Norbert Weiner's father-son memoir *Ex-prodigy*, was not disinterested in my own intellectual development. We had a complex relationship, one not without conflict. I ended up at Swarthmore College, and graduated from that distinguished liberal arts college with a major in mathematics, and minors in philosophy and literature. The path of least family resistance led me to graduate school in mathematics, and eventually to a degree in applied mathematics with a concentration in economics, as I wrote a dissertation on the applications of certain ideas in stochastic differential equations to general equilibrium systems in economics. I was but one of many mathematicians who crossed over to economics in those years, and I address these matters in chapters 7 and 8.

This book is thus associated with my family history, as it produced in me a life-long concern with connections across disciplines, in particular the interconnection of mathematics and economics. In larger or smaller measure, all of my work has played on this theme. As James Olney once wrote, quoting Paul Valery, "All theory is autobiography." Consequently as I have grown and established my own perspectives on my life and my work, the volume here represents a complex set of projects, or a life, whichever you the reader prefer to read.

In terms of stories already told by others, and to indicate my story's connection to theirs, let me recognize here some important explorations of twentieth-century mathematics and economics. The remarkable book by Bruna Ingrao and Georgio Israel, *The Invisible Hand* (1989), sets a historiographic standard. Their history of the interconnection of mathematics and economics emphasized general equilibrium analysis and the introduction of various analytical techniques into economic theory. Their metanarrative, however, concerned the capture of economic analysis by ideas alien to the foundations of that economic analysis. Their story, critical of modern economics, ultimately is one of how economists were deflected from their appropriate concerns by a mathematics that did not permit continued expression of a number of important ideas.

Philip Mirowski's 1989 volume, *More Heat than Light*, and his most recent book *Machine Dreams: Economics Becomes a Cyborg Science* (2001), are both totalizing narratives that frame the development of economics, a fortiori mathematical economics, over the twentieth century. Mirowski's story is one in which economics as science develops from late-nineteenth-century rational mechanics, and energetics, then moves over the first forty years of the twentieth century along the conflicting tracks set out in that turn of the century set of understandings, and finally is transformed by its connection to the cyborg sciences as they developed during and after World War II. For Mirowski, the story of mathematics and economics is a story of physics and economics, and later information science and economics. The meticulous archival research, oral history, and political and social history that shape Mirowski's narrative have produced a new historiography for economics. Nevertheless for Mirowski, "in order to truly understand the impasse of neoclassical economic theory, we must appreciate that the importation of physical metaphors into the economic sphere has been relentless, remorseless, and unremitting in the history of economic thought. Simple extrapolation of this trend suggests that it will continue with or without the blessing or imprimatur of orthodox neoclassical economic theory" (1989, 395).

In contrast to Ingrao and Israel, and Mirowski, I am not sympathetic to using history in order to criticize the discipline of economics. It is not that I have no beliefs about the strengths or weaknesses of particular lines of economic analysis. It is rather that, as a historian, both my interests and my task are different from that of an economist who wishes to argue with other economists about current economic analysis and policy.

Closer in spirit to my own project is Mary Morgan's recent work (Morgan 1999, Morgan and Morrison 1999) on the history of modeling in economics. Economics at the end of the twentieth century is a discipline that concerns itself with models, not theories, so how did this happen and what does it mean? Morgan interrogates the history of models in ways that I find most congenial. There is not one story, nor one set of meanings, but a variety of different uses, implications, nuanced understandings, definitions, and modes of argument that define a concern with the history of modeling in economics. A philosopher asking, "What is a model really?," and creating distinctions among ways of modeling or thinking about modeling, is not of much help to Morgan in her project. Instead, for Morgan, modeling is embedded in practice, in the craft of the economist, and thus she grounds her investigation in the history of practice and craft rather than in the history of economic theory.

So too with mathematics and economics. Doing mathematics and doing economics are two practices that developed within different communities. When those practices meet across community boundaries, we have an opportunity to construct histories less totalizing, less critical of practice, and more accepting of difference, change, and resistance to change. For beliefs do change, and knowledge changes, and beliefs and knowledge are mutually stabilizing. Historians of economics may usefully attend to the beliefs of economists, the nature of the stabilization of those beliefs into knowledge claims, the mechanisms by which such changes occurred, and the connections of those beliefs with other constellations of beliefs that themselves changed. The goal is to produce a richly interesting narrative integration of economic, social, and political history; economic and social policy; economic theory; the natural and social sciences; the rhetoric of economics; and the history and philosophy of economic thought.

We are a long way from that goal. My own project here is in fact a more limited one of producing a narrative integration of some history of mathematics with some history of economics, and thus telling a story of the development of economic analysis in the twentieth century.[1] This of course requires attending to the history of mathematics.[2]

The history that follows is not in the service of a larger set of claims that "this is the real story of economics in the twentieth century" or "economics in the twentieth century went off the path when . . ." The history that follows teaches no overarching lesson except to those who bring lessons to the history. Interwriting the history of mathematics and the history of economics of the twentieth century provides a framework for talking about what has happened, a framework for understanding where we are, and a perspective from the past on what alternative futures might be like. If the story enchants, delights, and raises the reader's empathetic understanding of the past, it will have done its job.

1   Burn the Mathematics (Tripos)

[Problem:] to analyse the motion of a smooth flat coin rolling inside the rough surface of a hollow ellipsoid balanced on the back of a hemispherical tortoise ambling at constant speed straight up a hill of uniform gradient on Saturn.
—I. Grattan-Guinness, *The Norton History of Mathematical Sciences*

Economists believe that the last third of the nineteenth century played a pivotal role in the evolution of their modern discipline. "The Marginalist Revolution," with its introduction of *homo oeconomicus* making consumption decisions at the margin, reshaped economics into a modern science. The controversies over the scientific status of economics were quite alive at the end of the nineteenth century as the German Historical School, American Institutionalists, the Austrian School, and others contested the nature and limits of economic science. The best way to do economics was in fact an open question in 1900. In this chapter, I shall argue that this contest shadowed similar contests in mathematics and physics, and its resolution was to be shaped by the resolutions that eventually stabilized those other fields of inquiry.

Similar to economists' beliefs about their own modern disciplinary origins, one of the central tropes of the history of mathematics concerns the crises in mathematics and physics toward the end of the nineteenth century that induced physicists and mathematicians to reconstruct their disciplines in the twentieth century. For mathematics, the crisis is often understood to have concerned the foundations of mathematics. There were three major threads: 1) the foundations of geometry, specifically the failures of Euclidean geometry to domesticate the non-Euclidean geometries; 2) the failures of set theory made manifest through Georg Cantor's new ideas on "infinity" (i.e., transfinite cardinals and the continuum of real numbers); and 3) paradoxes in the foundations of arithmetic and logic, associated with Gottlieb Frege and Guiseppe Peano. It is usually assumed that the response to these problems left the community of mathematicians unsure of what was right and proper and true and lasting in mathematics. For example, one popular exposition tells us that:

> Against the background of steady progress in the great scientific centers of England, France, Germany, Italy, and Russia, three sizzling developments in the last quarter of the nineteenth century prepared the ground for the massive explosion of new ideas in pure mathematics at the beginning the twentieth century: the creation (basically, single-handed) of the theory of infinite sets by George Cantor (1845–1918); Felix Klein's (1849–1925) announcement in 1872 of the Erlanger Program which proposed geometry as a discipline concerned with the study of an abstract object invariant under given transformation groups; the appearance in 1899 of *Grundlagen der Geometrie* by David Hilbert (1862–1943) axiomatizing Euclidean geometry. . . . All three came from Germany. They brought about a fundamental change both in the position of mathematics among other disciplines of knowledge, and the way mathematicians think about themselves. The aftershocks lasted well into the 1930s and beyond. . . . As a result, mathematics broke away from the body of natural sciences. (Woyczynski 1996, 107–8)

In the popular imagination however, even more critical was the failure of physics, particularly rational mechanics, to deal with the new problems raised by thermodynamics, quanta, and relativity. This led to a crisis in physics and, a fortiori, mathematical physics. That is, the kinds of nineteenth-century mathematics based on differential equations, and both quantitative and qualitative properties of dynamical systems, were fundamentally linked to the problems in mechanics. If the deterministic mechanical mode of physical argumentation was to be replaced by an alternative physical theory, some established areas of mathematics were no longer connected to a generally accepted physical model.

With Plank and Einstein there was a birth of a new physics: statistical mechanics, quantum mechanics, and relativity theory were to force physicists to think in terms of new models of the universe both large and small. Modeling the new physics required a new mathematics, mathematics based less on deterministic dynamical systems and more on statistical argumentation and algebra. Consequently, mathematical physics was to link up with newer mathematical ideas in algebra (e.g., group theory) and probability theory (e.g., measure theory), as mathematicians took up the challenge to work on mathematical ideas that could facilitate an understanding of the world.

Just as the objects of the physical world appeared changed—gone were billiard balls, newly present were quanta—the universe of mathematical objects changed. Transfinite sets and new geometries, together with a recognition that the paradoxes of set theory and logic were intertwined, led mathematicians early in the twentieth century to seek new foundations for their subject based on axiomatization, and formal modeling of the foundations of set theory, logic, and arithmetic. By the 1920s and 1930s mathematics was to become clear and coherent again after the "foundations crisis" of the turn of the century. In particular, it appears to be an established part of the general history of twentieth-century science that the problems, paradoxes, and confusions of turn-of-the-century mathematics were to be resolved by reconceptualizing the fundamental objects of mathematics just as physics had reframed the building blocks of the natural world.[1]

Whether or not one accepts this history of crises, looking first at mathematical work done before 1900, then at work done in the 1920s and 1930s, and finally at work done in the 1950s, it is clear that the mathematical landscape had been transformed. It is not our task to construct the history of those transformations, although in chapter 3 we shall look closely at competing histories of those changes in mathematics. Instead, as our concern is the transformation of economics, we need to attend to the changing features of the mathematical landscape as a background against which we might understand how economics was reshaped, over the first two thirds of the twentieth century, as a mathematical discipline.

Cambridge Mathematics

Since many English-speaking economists of the twentieth century trace their professional genealogy back to Alfred Marshall's *Principles of Econom-*

ics, let us enter the world of Marshall's Cambridge University, in England. As this university was the intellectual home of many of the English economists, it is an appropriate place to begin looking at the images of mathematical thought.

> Early in the nineteenth century the loose Georgian approach gave way to a somewhat more rigorous academic structure. In a period of rising enrollments, examinations began to play an increasingly important role at both Cambridge and Oxford. At Oxford, the examinations were in classics, which focus was justified as a way of broadening young minds rather than as imparting specialized knowledge. The same kind of rationale was applied at Cambridge, where, however, the central examination [called the Tripos] was in mathematics. Until the middle of the century one needed a pass on this examination in order even to take the parallel examination in the classics. Even though the Tripos became more and more mathematically demanding, the justification for requiring that the students study for it continued to be broadly humanistic rather than specific or professional. Throughout the century the center of England's mathematical education pursued the subject as a way to help students become fully formed human beings. (Richards 1991, 307–8)

The Tripos was a final set of examinations given to Cambridge students seeking a degree.[2] The name may derive from the medieval three-legged stool on which the candidate sat while being examined (Ball 1889), or it may have its origin in the Scholastic "Trivium" of grammar, logic, and rhetoric. In customary usage, one speaks of each particular Tripos, e.g., Natural Sciences, as defining a major field of study in which one could receive a Cambridge degree.

What did late-nineteenth-century English economists themselves study? The answer is not usually known except to specialists of the historical period, but in Cambridge, specifically, they usually studied mathematics. As Richards points out,

> Before 1848, the Tripos was an undifferentiated six-day [mathematics] examination. In the reform of 1848 it was lengthened to eight days and divided into two parts. The first three days were designed to cover the material essential for anyone to receive an ordinary degree. . . .

[After 1851,] when the Moral Sciences Tripos and Natural Sciences Tripos were added, students could attempt to receive honors on any Tripos after taking only the first part of the Mathematical Tripos. Thus, until the end of the century, the first part of the Mathematical Tripos remained the solid core of the education of any Cambridge graduate. (Richards 1988, 40)

What this meant was not only that one's undergraduate career, at England's premier university, was spent in large part studying mathematics, but in fact one studied a particular kind of mathematics. The Part I Tripos exam tested

the portions of Euclid usually read; Arithmetic; parts of Algebra, embracing the Binomial Theorem and the Principle of Logarithms; Plane Trigonometry, so far as to include the solution of Triangles; Conic Sections, treated geometrically; the elementary parts of Statics and Dynamics, treated without the Differential Calculus; the First Three Sections of Newton, the Propositions to be proved in Newton's manner; the elementary parts of Hydrostatics, without the Differential Calculus; the simpler propositions of Optics, treated geometrically, the parts of Astronomy required for the explanation of the more simple phenomena, without calculation. (*Report of the Examination Board for 1849*, as quoted in Richards 1988, 40–41)

The curious feature of this program is the emphasis on what is now thought of as "applied mathematics," actually rational mechanics. This point is better understood when one considers the history of the training of the mathematical examiners themselves, for in the years since Newton, up to the early years of the twentieth century, English mathematics stood apart from Continental traditions in the field of Analysis. The Tripos defined the concerns of English mathematics in a fundamental manner. The examination itself, but not mathematics that was important in the eyes of the best European mathematicians, defined the concerns of the students and the program. Indeed the very best Cambridge mathematicians, men like J. J. Sylvester, and Arthur Cayley, gave lectures that no students, or very few students, ever attended. Why should they have attended since that material was never going to appear on any examination? Instead, mathematical coaches like Hopkins and Routh prepared the students for the Tripos. Coaches arranged students in small classes of not more than ten.

> Each class was taken three times a week, on alternate days during the eight weeks of each of the three terms and the six or seven weeks of the Long Vacation; and in all there were ten terms and three Long Vacations in the full undergraduate course. Each class was for one hour exactly.... The topics were treated, not in connection with the general underlying principles that might characterize the subject, but in a way that the student should frame his answer in the examination.... At the end of the hour some questions ... [were handed out]; their solution had to be brought to the next lecture. (Forsyth 1935, 89)

Mathematics was thus defined, in England, by a set of tricks and details, based on Newton, which were linked to applied physics and mechanics, and which could be tested in a time-limited fashion. The function of the examination really was to provide a fixed ordering of the degree candidates. The top performer all the way down to the last found his place in the posted list of results. From Senior Wrangler (first place) to Second Wrangler (Marshall's place) to Third, etc., to Twelfth Wrangler (Keynes's place) down to Wooden Spoon (last passing grade), the order of finish of the Tripos defined one's options in the world of scholarship at least. Keynes, recall, did not get an academic appointment at Cambridge, his father's most fervent desire, because his Twelfth Wrangler position was simply not good enough that year. Consequently, he prepared for the Civil Service Examination instead of receiving a position as a Cambridge Fellow.

The "typical Tripos question, which has been parodied over and over again, was an unreal, often fantastically unreal, abstraction from the physical problem which had suggested it, whose sole object was to render it tractable to the candidates.... [The Tripos became] far and away the most difficult mathematical test that the world has ever known" (Roth 1971, 99, 97).

To make somewhat more specific the arguments that I seek to develop, consider the Mathematical Tripos questions for the year 1878. In this examination, we find questions like:

> viii. Describe the theory of Atwood's machine, and explain how it is used to verify the laws of motions.
>
> If the groove in the pulley in which the string runs be cut to that

depth at which it is found that the inertia of the pulley may be divided equally between the moving weights, and if Q be the weight required to be added to overcome the friction of the axle of the pulley when equal weights P are hung at the ends of the string, prove that an additional weight R will produce acceleration R divided by 2P + 2Q + R + W [all times] g where W is the weight of the pulley. . . .

2. The difference between the pressures at any two points of a homogeneous liquid at rest under gravity is proportional to the distance between the horizontal planes in which the points lie. A regular tetrahedron of thin metal, whose weight is equal to the weight of water it would contain, is emptied of water, and cut into two halves by a central section parallel to two opposite edges. If one half be held fast in any position, shew that the force required to draw away the other half from it will be the same, provided the centre of the tetrahedron be always in the same horizontal plane. (Glaisher 1879, 23)

The same examination also contained the following questions:

viii. Determine the initial motion of a rigid body which receives a given impulse; and find the screw round which it will begin to twist. A rough inelastic heavy ring rolls, with its plane vertical, down an incline plane, on which lie a series of pointed obstacles which are equal and at equal distances from each other, and which are sufficiently high to prevent the ring from touching the plane. If the rings start from rest from a position in which it is in contact with two obstacles, prove that its angular velocity as it leaves the (n + 1)th obstacle is given by

$$w^2 = 2\,\frac{g}{a}\,\sin i \sin \gamma \cos^4 \gamma\,\frac{1 - \cos^{4^n} \gamma}{1 - \cos^4 \gamma}$$

where a is the radius of the ring, i is the inclination of the plane to the horizon, and $2g/a$ is the angle which two adjacent obstacles subtend at the centre of the ring when it is in contact with both. (Ibid., 78)

Obtain the general equations of equilibrium of an elastic plate of small thickness, under given forces. [Next, assume a] thin uniform spherical shell of isotropic material, whose weight may be neglected, is made to perform vibrations in the direction of the radius, symmetrical about a diameter. Shew how they may be found. (Ibid., 211)

1. If the orbit in which a body moves revolves around the centre of force with an angular velocity which always bears a fixed ratio to that of the body's; prove, by Newton's method, that the body may be made to move in the revolving orbit in the same manner as the orbit at rest by the action of the force tending to the same centre. (Ibid., 220)

From these sample questions we can see how contrived, unreal, and even bizarre the Mathematical Tripos examination had become by the late nineteenth century. They show us a Cambridge mathematics community insulated from the concerns of continental mathematicians. The caricature posed by Grattan-Guinness in this chapter's epigraph gives the game away. The Mathematical Tripos was associated with "the great period of Cambridge mathematical physics: Ferrers, Green, Stokes, Kelvin, Clerk Maxwell, G. H. Darwin, Rayleigh, Larmor, J. J. Thompson" (Roth 1971, 223). Nevertheless, the Tripos set out a specific view of mathematics, which retarded understanding of pure mathematics as a logical, or structural, discipline. And the Tripos at the end of the nineteenth century was maintained in part by the eminence of the Sadlerian Professor of Mathematics, the remarkable Andrew Russell Forsyth.

Of Forsyth, it has been said that he

had the misfortune to be born a hundred years too late; in his mathematical outlook and technique he was a man of the eighteenth century. . . . Forsyth looks backward to Lagrange rather than forward to Cauchy. . . . [His 1893 *Theory of Functions of a Complex Variable*] is magisterial, Johnsonian; the author's powers of assimilation are well-neigh incredible—and yet, strange to say, despite his intentions and his absorption of the material, he never comes within reach of comprehending what modern analysis is really about: indeed whole tracts of the book read as though they were written by Euler. (Roth 1971, 225)

I bring this material to the reader's attention, in a discussion about mathematical economics, to suggest that English mathematics was the antithesis of what we now think of as rigorous mathematics. To all intents and purposes, there was no pure mathematics done in England in the nineteenth century, for even Cayley and Sylvester were not, as were Cauchy, Riemann, Weierstrass, Klein, and Lie et al., concerned with foundational issues of analysis. Modern mathematical ideas in England, as shared concerns of the larger world mathematical community, made their appearance with Hardy and Littlewood in the second decade of the twentieth century. The English mathematics studied by Second Wrangler Marshall, and Twelfth Wrangler Keynes, was by contrast a melange of applied physics, thermodynamics, optics, geometry, etc. It was as far from modern ideas of rigor as William Morris was from Piet Mondrian.

But rigorous it was nonetheless, as rigor was then understood to mean "based on a substrate of physical reasoning." The opposite of rigorous was unconstrained, as with a mathematical argument unconstrained by instantiation in a physical/natural science model. As Giorgio Israel (1981) has brilliantly argued, late-nineteenth-century mathematics considered "rigor" and "axiomatization" antithetical, whereas those two notions are virtually indistinguishable in mathematics of the late twentieth century.

The modern claim that Marshall did not provide a "rigorous axiomatic mathematical foundation" for economic theory is hardly surprising, for he would not have been able to comprehend our current idea of the nature and meaning of that phrase, in his vocabulary an oxymoron. Thus to note the lack of a formal/axiomatized mathematical economics, which usually means economics written in English, before the twentieth century, is to attend to the peculiarities of the Cambridge Tripos, which instantiated the late-nineteenth-century English idea of rigorous mathematical argumentation. To risk being repetitive, I cannot emphasize too strongly that although all mathematics, at least through much of the nineteenth century, required connected physical reasoning to be considered rigorous, by the end of the century this link was being broken in nearly all European countries except England. The cause of the backwardness of English mathematics was the backward-looking Tripos examination. Its importance to the institutional structure, developed to define a fixed order of merit in the "superior" English education, supported this rigidity.

Alfred Marshall

In Peter Groenewegen's biography of Alfred Marshall (Groenewegen 1995), we are presented with a detailed picture of the role that the Mathematical Tripos played in Alfred Marshall's own education, and how his mature subsequent work was intertwined with that most peculiar Cambridge institution. Marshall took the examination in its post-1848 form: "In the first three days six elementary papers were attempted. These decided whether a person was eligible to sit for the advanced, second part of the tripos to be examined over five days and ten further papers, following a week's interval after the examinations of the first part" (Groenewegen 1995, 80).

Groenewegen presents what Marshall was responsible for in preparing for those examinations. Among the usual parts of algebra, trigonometry, and conic sections, there are elementary portions of statics and dynamics, the latter treated without differential calculus; the first second and third sections of Newton's *Principia* with propositions "to be proved in Newton's manner"; elementary parts of hydrostatics, optics, and astronomy were also required (82). Second year required more work in calculus, differential equations, statics, and dynamics. Finally, the student moved to solid geometry, hydrostatics, dynamics, and optics at a more advanced level. Groenewegen notes that the subject matter emphasized "continuous application by the student through term and vacation" (83). His own discussion of the Tripos focuses almost exclusively on the need for coaching, rapid ability to problem-solve, and the need to develop the mastery of various set pieces that examiners tended to put on the examinations year after year.

Groenewegen quotes Arthur Berry's 1912 picture of the Tripos and Berry's view that

> the good mathematician who would naturally at this stage of his career have a bent toward certain departments of mathematics was much discouraged from any kind of specialization. The pure mathematician inclined to pursue the study of higher analysis would be checked by the necessity of being able to answer questions on geometrical optics. . . . In the examination . . . no place is assigned to any kind of original research. The examinations tested knowledge in that limited form of originality which consisted in applying knowledge very rapidly to such application of theory as could take the form of examination questions. . . . Another serious defect . . . is the almost complete divorce between mathematics and experimental physics. (Berry 1912, as quoted in Groenewegen 1995, 86)

In Sir J. J. Thompson's autobiography, he recalled his own experience in the Mathematical Tripos of January 1880:

[T]he examination for the Mathematical Tripos when I sat for it in January 1880 was a arduous, anxious and a very uncomfortable experience. It was held in the depth of winter in the Senate House, a room in which there were no heating appliances of any kind. It certainly was horribly cold, though the ink did not freeze as it is reported to have once done.

The examination was divided into two periods: the first lasted for four days. . . . At the end of the fourth day there was an interval of ten days in which the examiners drew up a list of those who, by their performance on the first three days, had acquitted themselves so as to deserve mathematical honours. These, and these only, could take the second part of the Tripos, which lasted for five days beginning on the Monday next but one after the beginning of the first four days. (56–57)

After discussing some of the material that appeared on the examination, Thompson recalls that

it was of great importance that the student should make no slips. . . . An error in arithmetic or a wrong sign in a piece of algebra, would involve going through the work again, and the loss of time when there was no time to lose. Accuracy in manipulation was perhaps the most important condition in this part of the examination . . . [the] qualities, having one's knowledge at one's finger-ends, concentration, accuracy and mobility owe their importance to the examination being competitive, to there being an order of merit, to our having to gallop all the way to have a chance of winning. (58)

Thompson concludes by arguing that the Tripos was not a useful rite of passage for all students:

The Tripos . . . was, in my opinion, a very good examination for the better men. It was, however, a very bad one for the majority who had not exceptional mathematical ability. Many of these men could cope with the more elementary subjects and benefit by studying them, but with these they had shot their bolt; they found the higher subjects beyond them and the time they spent over them wasted. (60)

As further evidence in support of Thompson's point, we have autobiographical notes, published in 1916, by Edward Carpenter who was 10th

Wrangler around 1870, and who was to hold a Mathematics Fellowship (to tutor astronomy) at Trinity Hall.[3] Carpenter (who also provided us in these notes with a memoir of F. D. Maurice) memorialized the Mathematical Tripos in this fashion:

> In coming up to Cambridge it had never occurred to me at the outset to go for an honours degree; my opinion of the University was too high for that. But after a term or two the tutor to my surprise seriously recommended me to read for the mathematical tripos. I was of course frightfully behind hand in my subjects, but I took a private coach, went through the routine of cram, and ultimately obtained a Fellowship.
>
> Mathematics interested me and I read them with a good deal of pleasure—but I have sometimes regretted that three years of my life should have been—as far as study was concerned—nearly entirely absorbed by so special and on the whole so unfruitful a subject. I think that every boy (and girl) ought to learn some Geometry and Mechanics; without these the mind lacks form and definiteness and its grip on the external world is not as strong as it should be; but the higher mathematics (certainly as they are read at Cambridge) are for the most part mere gymnastic exercise unapplied to actual life and facts, and easily liable to be unhealthy, as all such exercises are.
>
> After my degree, though retaining a certain general interest in the subject, I never again opened a mathematical book with the intention of seriously pursuing its study. (Carpenter 1916, 48–49)

In the same general period as that described by Thompson and Carpenter, Marshall emerged from detailed study and preparation with an outstanding result as Second Wrangler. It is noteworthy that Marshall did not enter a prize examination for the Smith Prize, which required some ability at original mathematical research. Groenewegen concludes his own discussion by attempting to assess "how good a mathematician did the Tripos make him?" He offers the assessment that "Alfred Marshall was to apply his mathematics to economics with care, with caution, and with a considerable degree of skill, a benefit from his Tripos experience and undergraduate period which should not be underrated though he preferred geometry more for this role than the terse language of algebra and the calculus" (94).

In summary, Groenewegen quotes John Whitaker (1975, 4–5) who argued that "despite an earlier penchant for Euclid, there is from the first an awkwardness and hesitancy about Marshall's efforts at mathematical economics that argues against him ever having breathed wholly freely on the pinnacles of abstraction. Both Jevons and Edgeworth seemed to have dwelt more comfortably in the realm of abstract logic, despite their inferiority to Marshall in mathematical training."[4]

There is, however, somewhat more to be said. Whitaker, for instance, goes on to say that "the common view [among economists] of Marshall as a mathematical giant who exercised great self-restraint in resisting for economics' sake the natural bent of his own mind, may have become exaggerated." However, Whitaker's and Groenewegen's discussions presuppose that mathematics is a monolith, although the idea of "being a mathematician" does not have a set of stable referents. Marshall himself, as is well known, was both a supporter of and an opponent of mathematical ideas in economics. This ambivalence certainly has it roots, as Groenewegen argues, in his own mathematical training as defined by preparation for the Tripos. Mathematics for Marshall was a competition, a venue for getting prizes and awards and topping the field. That he failed to take examinations that tested for mathematical originality is evidence of his own movement away from a mathematical career when he "only" managed to achieve the position of Second Wrangler. However, ambivalence carried over from his youth is not the entire story of Marshall's later comments on mathematical economics.

Of Marshall's support for work, and workers, in mathematical economics we have not only the evidence of his major book, but a wide range of supporting documents written in aid of the careers of those who pushed economics in that direction. An early piece is a note Marshall sent to Edgeworth on 8 February 1880, concerning Edgeworth's book *New and Old Methods of Ethics:* "I have now nearly read all of the book you sent me and am extremely delighted by many things in it. There seems to be a very close agreement between us as to the promise of mathematics in the sciences that relate to man's action. As to the interpretation of the utilitarian dogma, I think you have made a great advance. But I have still a hankering after a mode of exposition in which the dynamical character of the problem is made more obvious, which may in fact represent the central notion of happiness as a process rather than a statical condition" (as quoted in Whitaker 1996, 401).

In contrast, we have the famous letter to Arthur Bowley of 27 February 1906:

> But I know I had a growing feeling in the later years of my work at the subject that a good mathematical theorem dealing with economic hypothesis was very unlikely to be good economics: and I went more and more on the rules—(1) use mathematics as a short hand language, rather than as an engine of inquiry. (2) Keep to them till you have done. (3) Translate into English. (4) Then illustrate by examples that are important in real life. (5) Burn the mathematics. (6) If you can't succeed in four, burn three. This last I did often. . . . I think you should do all you can to prevent people from using mathematics in cases in which the English language is as short as the mathematical. (Groenewegen 1995, 413)

Groenewegen presents this letter by noting that "[Schumpeter] surmised that this practice reflected Marshall's peculiar ambition to be read by businessmen" (ibid.). However, Groenewegen suggests that "most crucial to the decision was Marshall's growing realization of the dangers in pursuing the logical consequences of mathematical reasoning in economics to the limit. . . . An economist's 'greed' for facts was an essential countervailing force to the thrill of the chase mathematical reasoning provided, if contact with reality of that economics were to be preserved" (ibid.). Groenewegen's Marshall appears in this interpretation to take on the opinions of the Joan Robinson Cambridge-era scholars for whom the late-twentieth-century world of mathematical economics was a wrong turn.

I think there is a more compelling explanation for this ambivalence, although it is an explanation that takes us outside Marshall himself. By the time of Marshall's writing to Bowley, we have an emergent mathematical economics with the works of Pareto, Panteloni, and others. Cournot, even with his mistake (in Marshall's view) concerning increasing returns, was beginning to be read, and Irving Fisher's book was not only in print but widely praised. These books reflected a mathematical sophistication and use of mathematics in essentially new ways. For a product of the old Tripos like Marshall, who grew up thinking of mathematics as concerned with deriving certain conclusions from geometric arguments, having as a model the memory of problem-solving set pieces of Newtonian mechanics by Newton's own (Euclidean) geometric methods under duress, this new way of using mathematics might have been discomforting. The point is that for Marshall, his image of what mathematics was, and how it was to be done, and especially how it was to be applied to problems, was forged by the Mathematical Tripos of his Cambridge student years, and his preprofessorial days there.

Groenewegan remarks, "The appreciation of mathematical knowledge as necessary and inevitable truth, derived axiomatically, was an aspect of Cambridge mathematical training which justified its pre-eminence in the university honours syllabus, combined as it was with the methods by which such truths could be mastered. This was a point stressed by Whewell in his defense of the value of mathematical specialization. A high wrangler in particular would have been heavily imbued by this specialized feature of mathematical knowledge" (116).[5] This point made by Groenewegen is extremely important. That is, for Marshall mathematics was indeed part of the Whewell program (Henderson 1996) whereby it served as an exemplar of the path to truth, to constructing indubitably true arguments. This is why it had such a central place in the early Victorian Cambridge educational process. However, as Richards has shown, that role was undermined by the first set of crises in mathematics in the nineteenth century that resulted from the construction of non-Euclidean geometries. Consequently, mathematics, particularly a mathematics based on Euclidean geometry and Newtonian mechanics approached through Euclidean geometry, was no longer a sure path to truth. No longer was mathematical knowledge a model for secure knowledge. The image of mathematics that Marshall had grown up with was no longer sustainable.[6]

That change in mathematics, based on a new conception of what mathematical truth might mean, occurred over the second third of the nineteenth century, and was well-incorporated in the Continental tradition in mathematics. That is, outside England there was a change in mathematics between the time of Whewell's defense of mathematics in the educational process, the time of Marshall's student days, and Marshall's later time as Professor of Political Economy. Whewell's mathematical era was not Marshall's. The emergence of non-Euclidean geometries had made Whewell's argument about axiomatics, and inevitable truth, ring hollow long before 1906 and Marshall's letter to Bowley. In the time of the new geometries, the difficulty of linking mathematical truth to a particular (Euclidean) geometry produced a crisis of confidence for Victorian educational practice, a point well-documented (Richards 1988). It in fact was this crisis that prepared the late Victorian mind for the new idea that mathematical rigor had to be associated with physical argumentation. Moreover, as we shall see, it was this image of mathematics in science that was to shape the concerns of individuals like Edgeworth and Pareto.

But by century's end the images of, and styles of doing, mathematics were to change yet again in response to Klein's studies of geometry, Cantor's set theory, and the new challenges to Newtonian mechanics arising in physics. As Continental influences were finally intruding upon the insular world of English mathematics, Marshall was caught. His image of mathematics was formed by the early Victorian Mathematical Tripos of simple geometry, the drawing of cord segments and conic sections, simple statics, dynamics, and the like.[7] His conception of mathematics was incompatible with either the late-nineteenth-century mathematics of physical-model-based analysis, or that which was to supplant it in turn, the early-twentieth-century move to axiomatics and mathematical-model-based analysis. The former shift would have required a measurement-based mathematical economics, while the latter would have required a move away from the study of "mankind in the ordinary business of life."

The paradox of Second Wrangler Marshall growing increasingly suspicious of mathematics has been seen as a problem for historians of economics from the perspective of an unchanging mathematics and a changing Marshall, a minor *Das Alfred Marshall Problem:* was Marshall's view of mathematics continuous over his life, or did he change his mind about the role of mathematics in economics? If the latter, the historian of economics then needs some explanation for Marshall's changes. What I am suggesting is an inversion of the usual picture. I submit that there is considerable explanatory power in the suggestion that Marshall's image of mathematics was formed in his own Mathematical Tripos experience and was generally unchanged through his lifetime. Marshall's "advice" to Bowley was given by a sixty-three-year-old scholar close to retirement; I am reminded of the peroration in Keynes's *General Theory* in which he notes that "in the field of economic and political philosophy there are not many who are influenced by new theories after they are 25 or 30 years of age" (Keynes 1936, 383–84). So too for mathematics.

Felix Klein

In order to get a fuller grasp of what the alternative images of mathematics could be, circa 1900, let me now turn aside from economists, and instead ask how mathematicians were representing themselves and their enterprise. What were they saying about the right way, the best way, to think about the nature and role of mathematics? What, in other words, was the context, in the community of mathematicians, for the views held by economists on the role of mathematics in economics?

When Felix Klein visited the United States in 1893 to deliver the Evanston lectures at Northwestern University, in conjunction with the World's Columbian Exposition in Chicago, he was perhaps the most important "American" mathematician, although he was German and his home was Göttingen. As is well-documented (Parshall and Rowe 1994), Klein had been the Ph.D. thesis adviser to almost an entire generation of American mathematicians whom he trained in Germany, and so his invitation to come to the United Sates to provide a survey of mathematics was entirely appropriate. Of special interest to us is his sixth lecture, delivered 2 September 1893, titled "On the Mathematical Character of Space-Intuition and the Relation of Pure Mathematics to the Applied Sciences" (Klein 1894).

He began by distinguishing naive intuition and refined intuition. That is, axioms refine the basic ideas that naive intuition constructs in geometry, while refined intuition in Klein's view is not properly intuition at all "but arises through the logical development from axioms considered as perfectly exact" (42). Following this discussion of intuition and how it is connected to the development of axioms for particular subjects, specifically geometry, Klein states that he himself believes that one never gets to the fully axiomatized nonintuitive state: "I am of the opinion that, certainly, for the purposes of research it is always necessary to combine the intuition with the axioms" (45).

Klein's lecture then moves to an extraordinarily interesting set of observations on the role of mathematics in the applied sciences. "From the point of view of pure mathematical science I should lay particular stress on the historic value in applied sciences as an aid to discovering new truths in mathematics. Thus I have shown . . . that the Abelian integrals can best be understood and illustrated by considering electric currents on closed surfaces . . . and so on" (46). In other words, the applied fields themselves nurture mathematics by providing a source for problems, and ways of thinking about (models for) mathematical structures. The connections between mathematics and the sciences are not unidirectional, but rather flow both ways.

He continues by arguing that

> I believe that the more or less close relation of any applied science to mathematics might be characterized by the degree of exactness obtained, or obtainable, in its numerical results. Indeed, rough classification of these sciences could be based simply on the number of significant figures on average in each. Astronomy (and some branches of physics) would here take the first rank; the number of significant figures attained may here be placed as high as seven . . . chemistry would probably be found at the other end of the scale, since in this science rarely more than two or three significant figures can be relied upon. . . . The ordinary mathematical treatment of any applied science substitutes exact axioms for the approximate results of experience, and deduces from these axioms the rigid mathematical conclusions. In applying this method it must not be forgotten that mathematical developments transcending the limit of exactness of the science are of no practical value. . . . Thus, while the astronomer can put to good use a wide range of mathematical theory, the chemist is only just beginning to apply the first derivative . . . for second derivatives he does not seem to have found any use as yet. (46–47)

This is a long way from Whewell. Mathematics is not really of much fundamental use in a science unless that science is able to constitute its basic concepts with "exact axioms" and precise numerical results. To truly imitate physics, a science of political economy would need to have measurable quantities of its conceptual building blocks, and ways of measuring its "results." The prerequisite for having a mathematical science is to have exact measurements in that science. This is not the image of mathematics aiding scientists to achieve "clear reasoning to certain conclusions."

Klein ends his lecture with some observations on the educational implications of these connections between mathematics and applied sciences. "I am led to these remarks by the consciousness of the growing danger in the higher educational system of Germany,—the danger of a separation between abstract mathematical science and its scientific and technical applications. Such separation could only deplored; for it would necessarily be followed by shallowness on the side of the applied sciences, and by isolation on the part of pure mathematics" (50). The import is clear. The separation of mathematics from its applications would be hurtful to mathematics's claims for generality and applicability. Klein "was the geometer par excellence, a grand synthesizer, forever seeking out the visual element that bring a theory to life . . . [who at a first approximation] saw the burgeoning interest in abstract structures and axiomatics as a potential threat to the lifeblood of mathematics" (Rowe 1994, 188–89).

In Klein's image of mathematics, mathematical argumentation in any applied field requires quantitative argumentation in that field in order to ground the analysis. The idea that one can have a useful mathematical theory of X, where X could be astronomy, economics, or forestry, would appear then to require physical modeling, mechanical modeling, in the manner of a successful mathematical physics. The success of any applied mathematical field would then be linked to a reductionist argument of the form "the nearness of field X to physics is an indicator of likelihood of successfully producing a mathematical theory of X." It was not that Klein himself developed such a reductionist perspective, but rather that no alternative appeared to be viable for him. His call for the study of mathematics together with its applications is our first hint that there was an alternative image of mathematics developing in the community of mathematicians. At the end of the nineteenth century, Klein's vision looked backward to the successes of mathematics in physics. However, those successes themselves were to be questioned within a decade as Einstein and Planck called the reductionist mechanical program itself into question. Nevertheless, that change in the image of mathematics in its connection to science was to come a bit later, and while Klein was not its enemy, neither was he its champion.

Francis Ysidro Edgeworth

Let me continue this line of argument by considering some of the work of Francis Ysidro Edgeworth. In 1889 Edgeworth gave the opening address to section F of the British Association.[8] His talk, published in *Nature*, was titled "Points at Which Mathematical Reasoning is Applicable to Political Economy" (Edgeworth 1889). In it Edgeworth set himself the task of moving on from other "writers [who] seem to present what I may call the economical kernel of Jevons' theory divested of the mathematical shell in which it was originally enclosed; whereas my object is to consider the use of that shell—whether it is to be regarded as a protection or an encumbrance" (132). He argued that "our mathematical method rightly understood . . . is concerned with quantity, indeed, but not necessarily with number. It is not so much a political arithmetic as a sort of economical algebra, in which the problem is not to find x and y in terms of given quantities, but rather to discover loose quantitative relations of the form: x is greater or less than y, and increases or decreases with the increase of z" (133). Edgeworth quoted with approval Clerk Maxwell's view that the ideas as distinguished from the operations and methods of mathematics are what are important for mathematical physics, noting "algebra and geometry are to ordinary language in political economy somewhat as quaternions are to ordinary algebraic geometry in mathematical physics" (134).

Following a discussion of mutual dependence and equilibrium, Edgeworth notes that "the language of symbol and diagram is better suited than the popular terminology to express the general idea that all things are in flux, and that the fluxions are interdependent" (135). It is not that the mathematical theory can lead one to concrete or numerical deductions. "It may be doubted whether the direct use of mathematical formula extends into the region of concrete phenomena much below the height of abstraction to which Jevons has confined himself. However, the formulation of more complicated problems has still a negative use, as teaching the Socratic lesson that no exact science is attainable" (139).

Continuing in this vein, Edgeworth looks to other sciences for guidance:

> As compared with mathematical physics, the mathematical theory of political economy shows many deficiencies. First, there is the want of numerical data, which has been already noticed. . . . Much of our

reasoning is directed to the refutation of fallacies, and a great part of our science only raises us to the zero point of nescience from the negative position of error. . . . It is that in our subject, unlike physics, it is often not clear what is the prime factor, what elements may be omitted in a first approximation. . . . It will not be expected that from such materials any very elaborate piece of reasoning can be constructed. Accordingly another point of contrast with mathematical physics is the brevity of calculations. The whole difficulty is in the statement of our problems. The purely computative part of the work is inconsiderable. Scarcely has the powerful engine of symbolic logic been applied when the train of reasoning comes to a stop. . . . It follows that in economics, unlike physics, the use of symbols may perhaps be dispensed with by native intelligence. . . . The parsimony of symbols, which is often an elegance in the physicist, is a necessity for the economist. Indeed, it is tenable that our mathematical construction should be treated as a sort of scaffolding, to be removed when the edifice of science is completed. (143–46)

Edgeworth winds down his talk by stating that

You will perhaps come to the conclusion that the mathematical theory of political economy is a study much more important than many of the curious refinements which have occupied the ingenuity of scientific men; that as compared with the great part of logic and metaphysics it has an intimate relation to life and practice; that, as a means of discovering truth and in educational discipline, it is on a level with the more theoretical part of statistics; while it falls far short of that sort of pre-established harmony between the subject-matter and the reasoning which makes mathematical physics the most perfect type of applied science. (147–48)

Edgeworth's continuous referring to mathematical physics in this address echoes his 1881 *Mathematical Psychics* (Edgeworth 1985 [1881]). There, in his Appendix 1 "On unnumerical mathematics," Edgeworth considers both what some earlier writers thought would be the contribution of mathematics to economics, and what he himself believes to be the domain of the subject. Edgeworth considers the use of arbitrary or general functions to represent particular ideas, and the reasoning in making inferences about those functions, to be the appropriate domain of the subject. He suggests that this is precisely analogous to the situation in mathematical physics, and which makes possible an analogous theory of mathematical "psychics," or individual human behavior, a theory on which a scientific political economy could be based.

> The great theories relating to energy present abundantly mathematical reasoning about loose indefinite relations. Conservation of energy is implicated with such a relation, the mutual attraction of particles according to some function of the distance between them. . . . Peculiarly typical of psychics [a footnote compares "pleasure" and "energy"] are the great principles of maximum and minimum energy. That a system tends to its least potential energy, this principle affords us in innumerable instances a general idea of the system's position of rest; as in the very simple case of equilibrium being stable when the centre of gravity is as low as possible. Thus, without knowing the precise shape of a body, we may obtain a general idea of its position of equilibrium. [Moreover,] . . . upon analogous principles in statical electricity, we know that, if there be a given distribution of electricity over the conductors in a field, the strains throughout the dielectric is such that the potential energy of the whole system is a minimum. We may not know the precise form of the functions which express the distribution of electricity . . . yet it is something both tangible and promising to know mathematically that the potential energy is a minimum. That something is the type of what mathematical psychics have to teach. (68–70)

Edgeworth closes this discussion with similar remarks about physics and the calculus of variations. In sum, Edgeworth's mathematical economics is continuously referenced to mathematical physics. For Edgeworth, to apply mathematical reasoning to the science of political economy is to look to mathematical physics to see how mathematics can aid in the constructions of propositions within the science. Indeed, the view of a mathematics (in contrast to statistics) connected to the world is a result of the world's construction out of the disciplines that grew from natural philosophy. In the land of Newton, and in the century of science, how could it be otherwise?

I note here that Philip Mirowski would give a somewhat different reading to Edgeworth's role in the mathematization of economics. For Mirowski, Edgeworth came from a different tradition than did Marshall, and sought in psychophysics the model for economics as mathematical psychics.[9] That is, Mirowski has argued that the referent for Edgeworth, though it was at its root mathematics cum physics, was at one remove from mathematical physics, for the exact model for the use of mathematics in economics was Edgeworth's conception of the use of mathematics in the embryonic field of psychology, or as it was developing as "psychophysics":

> [There] is no evidence he ever wavered in his allegiance to utilitarian psychophysics to the day of his death; and likewise, his faith in physical analogy also persisted unabated. What changed, rather, was the world around him, rendering the untrammeled advocacy of his beliefs untenable. Partly, it was due to a shift in his guiding star, physics itself. The decade from 1895 to 1905 was perceived at the time as one of serious and profound upheaval, in which energetics itself was banished by such unexpected phenomena as radioactivity, and turned on its head by the novel theory of relativity. Ysidro's exemplar, the Maxwellian gas laws, no longer seemed so central to new trends in physics. (He did not live long enough to take note of how Planck's quantum arose from black-body thermodynamics.) There was also . . . the concomitant loss of interest in Fechnerian psychology, and its displacement by Freudian theories. There was, moreover, the vexation of Karl Pearson's positivist approach to science and statistics coming to dominate the British scene. (Mirowski 1994, 48–49)

With respect to our own developing argument, Edgeworth's writing presents us with an image of mathematics only partially tied to Whewell's older ideas of mathematics as a model for certain knowledge. Instead, he conceives mathematics as the intellectual structure in which physical reasoning may be developed. It is the underlying physical model that political economy needs to imitate, and the mode of imitation is the mathematical structure of that model. Yet, as Mirowski argues, that underlying physical model is itself beginning to fray at the end of the nineteenth century, with the challenges presented by Einstein and Planck. Edgeworth's image of mathematics was too wedded to an increasingly obsolescent physical model to allow his mathematical political economy to develop much further. He instead turned his attention in the new century more and more to statistics, both pure and applied, as the tool to unlock the secrets of human behavior.

Vito Volterra

Though we will be talking about Vito Volterra at greater length in the next chapter, it is appropriate in the current context to listen to this important mathematician specifically discuss the relationship of mathematics to economics at the beginning of the twentieth century. The occasion was Volterra's inaugural address as professor at the University of Rome (Volterra 1906a), which provided an opportunity for this gifted and distinguished international scientist to reflect on ways in which his own field, mathematics, could potentially enrich discourse and practice in the social sciences and biological sciences.[10]

> Among most mathematicians, however, comes the natural desire to direct their mind beyond the circle of pure mathematical analysis, to work toward comparing the success of different methods that it holds and to classify them based on applications in order to use its activity to perfect the most useful methods, reinforcing the weakest and creating the most powerful. Curiosity is the most intense about the sciences that math has ventured into most recently, I am, of course, speaking mainly about the biological and the social sciences. It is all the more intense because of the great desire to make sure that classical methods which have given such clean results in mechanical-physical sciences are likely to be transported with equal success into new and unexplored fields which are opening before it. (Volterra 1906b, 1–2)

Volterra reminds his listeners that

> mathematics only produces what you feed it and that the analysis adds nothing essential to the postulates that constitute the substance of any mathematical development. But it is no less accepted that mathematics is the most efficient way of gaining access to general laws. It is the surest guide that allows us both to imagine new hypotheses and to

improve those very postulates which are the basis of any study. They offer the most perfect means to test postulates and to transport from the abstract into the domain of reality. Nothing is better than calculus to exactly compare their furthest consequences with data based on observation and experience. (2)

Volterra pays his respect to the "otherness" of the mathematical community:

> By profession, mathematicians awkwardly separate themselves from the rest of the world by a barrier of symbols which give a mysterious aspect to their wild imaginings and to their works, to such an extent that those who are not initiated into the secrets of algebraic calculus sometimes have the illusion that their procedures are different in nature from those where ordinary reasoning dominates. . . . This shift of the signs of a pre-mathematic period to one where science tends to become mathematical, is characterized in the following way. The elements that it studies cease to be examined from a qualitative point of view and come to be examined from a quantitative one. As a result, in this transition, the definition which recalls the idea of elements in a rather vague form give way to definitions and principles which determine elements by indicating the means to measure them. (3–4)

> Therefore, first establish comments in a way that allows the introduction of measure, and from those measures discover laws, from those laws work back the hypothesis, then by means of analysis, deduced from the hypothesis a science which reasons in a rigourously logical manner about ideal beings, compare consequences to reality, reject or transform the recycled fundamental hypothesis as soon as a contradiction appears between the results of the calculation and the real world, and in this matter succeed in guessing in new facts and new analogies, or deduce once again from the present state what the past was and what the future will be. This is, quite briefly, how one can summarize the birth and evolution of a science which has a mathematical character. (5)

Volterra shows himself completely at ease with the emerging literature in mathematical economics. He provides a lucid discussion of *homo oecono-*

micus and Pareto's ophelimite and indifference curves. He cites Panteloni, Pareto, Irving Fisher, Barone, Jevons, Whewell, Cournot, Gossen, Walras, and even goes back to Ceva Giovanni of 1711. He notes that "once our researcher has examined the logical method employed in obtaining the conditions of economic equilibrium, he will recognize the reasoning which allows him to establish the principle of virtual labor. And when he finds himself faced with differential equations of economics, he feels the urge to apply methods of integration to them" (8).

Volterra then moves to a discussion of the biological sciences, noting work by Helmholtz, Weber, and what was coming to be called biometrics. Volterra makes the distinction between economics and biology to the effect that economics works mathematically at the level of analysis whereas biology works mathematically in the area of statistical analysis—methods of large numbers and of probability calculus. For economists, "these great leaps have allowed the creation of the branch of political economy that Descartes and Lagrange would probably have called analytic economics as an autonomous branch of science, and have also fostered the even more recent starts of biology in quantitative and statistical research" (14–15).

What we see for Volterra at the beginning of a new century is the enthusiasm of a mathematician, who himself had done significant work in mathematical physics, for the emerging theories in economics and biology. Volterra, with major contributions to the theory of integral equations, was a creative mathematician. Rigorous mathematics was founded on ideas of what we would now call applied mathematical theory. Mathematics was importantly tied to applications of analysis, and those applications themselves had to be structured to facilitate questions of measurement and prediction. At a time when Marshall was calling for restraint in the application of mathematics to economics, and deeply suspicious of attempts to measure economic concepts like utility, we find Volterra suggesting that the work done was both interesting and potentially quite grand. For Volterra, as for Klein, the need in a field like economics was for measurement. For Volterra, as for Edgeworth, concepts had to be developed that would allow exact calculations, for that was the route to a mathematical science like the physics that was the paradigmatic mathematical science. Underlying Volterra's image of mathematics in economics in 1900 then was a belief that it was possible for economics to develop in such a fashion.

Indeed there was not any other candidate perspective Volterra considered. Physics was his model for scientific certainty. That physics itself was losing its certitude went unremarked, because it was unnoticed.

Vilfredo Pareto

In a 1911 paper, the Italian economist Vilfredo Pareto, who followed Leon Walras as the professor of political economy at Lausanne, recapitulated several of his arguments from his earlier reputation-making *Manuel* and focused rather exclusively on the nature of mathematical economics. He began by stating "as in all studies concerned with the application of mathematics, we are faced with two quite distinct problems: (a) an exclusively mathematical problem, which derives the consequences of certain assumptions; (b) a problem of the adaptation of the assumptions and their theoretical consequences to concrete practical cases. It is with the first of these two problems that we shall chiefly deal in this article" (Pareto and Griffin 1955 [1911], 58). He further notes that "most of the objections which have been raised against the theories of mathematical economics are in reality objections to particular applications of those theories . . . [but] the mathematical problem which is the subject of our study can be stated in these terms: given the mathematical laws according to which certain individuals usually behave, determine the consequences of these laws" (ibid.).

Pareto is quite explicit in his linking individuals in economics with the particles studied in mechanics: "The position of the individuals concerned will be called an equilibrium position if it is such that, according to the given laws, the individuals can remain there indefinitely. As in mechanics, we shall have to consider stable equilibria and unstable equilibria" (ibid.).

Setting up supply-and-demand functions, and writing down the system of equations that result in equilibrium, Pareto remarks that

> all literary economics may be portrayed as attempts to solve, in everyday language and without the use of mathematics, this system of equations and other similar systems which we have for production. To this end, literary economists have tried, with the aid of more or less plausible devices, to reduce these systems of equations to one or two equa-

tions at the most; for that kind of problem was the only type which the state of their knowledge permitted them to grasp. (60)

Pareto goes on to discuss components of the general mathematical structure of economics that he has developed in his own work: the idea of index functions (or what were to be called indifference curves), maxima, constraints, etc.[11] After developing a great deal of the analytics of the subject, Pareto concludes "now the study of economics presents us with the following propositions: (a) given index functions and relationships, what are the equilibrium points? This is a proposition of pure economics and it is with questions of this kind that we have here been concerned" (101). The more than forty intervening pages really reflect a style and mode of argumentation. They present mathematical economics by *doing* mathematical economics. Nevertheless, Pareto follows the outline of concrete cases by remarking that

> a profound error, of which unfortunately certain mathematical economists are not innocent, is to imagine that mathematical economics can directly solve the problems of practical economics. This is not the case. Mathematical economics is only one of many parts which, united by synthesis, can provide a solution of practical problems. It bears the same relations to them as theoretical mechanics does to the problems of applied mechanics, as thermodynamics does to the practical studies of the construction and use of steam engines, as chemistry does to the practice of agriculture etc. The objections which have been raised against the study of mathematical economics are neither more nor less valid than those which have already been raised against the study of theoretical mechanics, thermodynamics and other similar sciences. (88–89)

Pareto's image of mathematics is much the same as Edgeworth's: mathematical argumentation instantiates basic propositions in mathematical logic and derives the implications of those basic propositions. It can lead to conclusions although those conclusions are generally qualitative not quantitative. They allow representation of arguments like "if an industry is organized along the following lines, what are the characteristics of the firms when that industry is in equilibrium?" Mathematical economics cannot instruct a manager, nor can its reasoning chains guide a finance minister. Mathematics, for Pareto, referred directly to mathematics as it was applied to physics, rather than the kind of empirical matters that concern engineers. The mathematics that Pareto himself had learned as a student were fashioned in this way, and so mathematical economics could take no other form.

Yet, Pareto points out, the mathematical argument is a formal argument, based on a structure of assumptions, definitions, and "laws" that work to move arguments to conclusions. The underlying physical model is not so much, as it was in Edgeworth, there to guide the argument along reasonable lines as it is to suggest how an argument can be made to work. I do not want to make too much of this point, but simply to call attention to the image of mathematics that is implied by this other perspective: one can have a mathematical model of a phenomenon without having a physical model of that phenomenon. For many mathematicians and scientists at the end of the nineteenth century, such a physically unconnected mathematical theory would be a paradigm for a nonrigorous theory, a way of doing mathematics "with the net down," as it were. The tension between these two views of mathematics would resolve itself in major changes in mathematics in the new century.

Remaking Mathematics for the New Century

Subsequent chapters will explore the interconnection of economic analysis with changing images of mathematics that took place in the twentieth century; specifically, how mathematicians thought about mathematics mediated economists' own thinking about mathematics and thus shaped how economists came to understand the enterprise of making economics more mathematical. By 1900 of course, arguments in favor of making economics a mathematical science had been circulating for decades. The calls to turn economics into a science, which grew out of the successes and prestige of science in many countries over much of the nineteenth century, gave way to a new understanding that for economics to take its place as the queen of the social sciences, it needed to emulate the queen of the sciences itself.[12] Subsequent chapters will take up, in different ways, exactly what this means. However, before proceeding to address such questions in a larger generality, let me reiterate how these issues would have appeared at the time to Volterra and Klein. How, as the mathematical landscape was being transformed around 1900, might these mathematicians have thought about a mathematical economics, and how would it be related to their views about mathematics itself?

Both Volterra and Klein, before century's end, shared a perspective that suggested that applications of mathematics to other disciplines (they specifically named economics and biology and physics) would depend upon accurate measurements and observations and modeling strategies in the applied disciplines. Such strategies would permit the use of mathematics, apparently the mathematics of deterministic dynamical systems connected to rational mechanical argumentation, to define a reasonable applied program. Such deterministic reductionism would be quite unremarked at century's end by mathematicians. These ideas would begin to change around that time, but they had not changed so clearly and distinctly that all mathematicians would have agreed that change had occurred. To an economist looking to mathematics then, one could construct an economic theory and base it on a structure of mathematical reasoning, which itself would be consistent and coherent and, as much as the word could be used, true.

It is this kind of reductionist argumentation that draws the critical attention of Philip Mirowski in *More Heat Than Light,* where he shows how that argument functioned in Leon Walras's attempts to connect his theory to the views of Poincaré, who by 1900 had begun to abandon such reductionist ideas. Contrasting Volterra with Poincaré, Ingrao and Israel (1990) argue that:

> Volterra followed Poincaré in regarding the relationship between theory and empirical data (and the connected issue of the measurability of the theory's basic magnitudes) as crucial, but his eager attitude was a far cry from that adopted by the French mathematician. This was due to their different views on the subject of applied mathematics and on the possibility of transferring the formal-explicative model of mathematical physics to other sectors. While both were aware of the crisis science was undergoing at the time, their reactions were different. Poincaré tackled the crucial themes of contemporary physics head-on;

Volterra ducked them and sought to consolidate the classical point-of-view by extending its field of application into other sectors. (163)

While Klein sought applicability of mathematics and looked to physics, Volterra worked at it in both economics and biology. Mirowski details the difficulties that Walras had with a Poincaré who was not persuaded by the reductionist perspective; we can understand Mirowski's argument in our framework to be that Walras was dealing with a mathematician less wedded to the past than Volterra. Indeed, once Walras became aware of Volterra's own perspective, he shared the good news with his own son, writing to him "you have grasped the importance of Volterra's article perfectly. As a skilled mathematician, he immediately recognizes that the revolution we have attempted and even accomplished in political and social economy is absolutely the same as those carried out by Descartes, Lagrange, Maxwell, and Helmholtz in geometry, mechanics, physics and physiology" (Ingrao and Israel 1990, 162).

Seen in this light, the pessimistic arguments of Marshall concerning mathematics appear directed at a particular kind of reductionism, a specific vision of the application of mathematics commonly understood in the years before the end of the nineteenth century, but an image of mathematics uncongenial to one brought up on Whewell's ideas about the role of mathematics.[13] We cannot construe that argument as disfavoring the notion of the applicability of mathematics itself: indeed our point has involved the inadmissibility of any such essentialist idea. This is why Marshall's comments about the need to "burn the mathematics" are so curious to a modern economist. Yet if the "burn" refers to Marshall's refusal to accept a rational mechanics reductionism because his Tripos-formed image of mathematical knowledge compels "applicable mathematics equals reduction to a Newtonian-Euclidean mechanical system," his comments in the letter to Bowley appear to make some sense. But then Marshall becomes even more impossible to reconstruct from the even later perspective on mathematics in which an economic problem can be interpreted in terms of a mathematical (as opposed to a specifically physical-mechanical) model. With our present-day image of mathematics, the letter to Bowley shows that Marshall must have "turned away" from his youthful optimism.

Thus, 1900 proves to be a difficult starting point for our discussion, for mathematics itself changes again early in the twentieth century. Not only do we find Marshall looking back to old Tripos days, but Edgeworth and Volterra and Pareto were arguing about the need for an empirically grounded economic science that could successfully employ mathematical ideas, while at the same time Klein was looking to keep the new mathematics of axiomatics at bay. All of them were writing about mathematics around 1900, but there were at least three images of mathematics tangled together in those discussions about the nature and role of mathematics in economics early in the twentieth century.

The nature of the changes from a mathematics grounded in truth-making, to a mathematics grounded in physical argumentation, to a mathematics shaped by what is referred to as mathematical formalism, are extremely complex and contentious issues in the history of mathematics. Nevertheless, as we reflect these issues back through economics, we need to see more precisely how they began to work themselves out in the practices of mathematicians doing economics. Let me now turn to a more comprehensive examination of the Italian whom we have just met who gave up on mathematical economics, and his American mathematical disciple who attempted to refashion it: Vito Volterra and Griffith Conrad Evans.

2 The Marginalization of Griffith C. Evans

The historian serves no one well by constructing a specious continuity between the present world and that which preceded it.—H. White, *The Tropics of Discourse*

It is a convention in the history of economics of the twentieth century to contrast a pluralistic interwar economics with a monolithic (neoclassical) postwar economics. There are two conflicting metanarratives in play with this idea: first, we have the triumphalism of disciplinary progress, a morality play in which economics finally fulfills its Jevonian promise and becomes scientific through the use of mathematics. In this view, good science displaces bad thinking, loose thinking, and the inappropriately varied argumentation of economists. We may think of Paul Samuelson as an exemplar of this way of constructing the history of this period (Samuelson 1987). Alternatively, we may see and thus construct the interwar period as one in which the healthy variety of economic thought was forced onto the procrustean bed of neoclassical theory: what emerged by the time of the neoclassical synthesis was an economics bereft of joy, intelligence, and humanity. We may think of modern Institutionalists, neo-Austrians, and Post Keynesians as exemplars of this perspective. In either case, the usual question asked is "Why was pluralism replaced by neoclassicism?" The two metanarratives condition the meta-answers, "So that Goodness would triumph" or alternatively, "So that Evil would triumph." In both cases however, it is apparently believed that mathematical theory "pushed out" non-mathematical theorizing in economics.

In this chapter I will not address the "why" question posed above. Instead I shall look closely at "what happened" in this period by reading Griffith Conrad Evans in conjunction with the mathematician Vito Volterra, whose image of mathematics, and mathematical life, was rather different from that of the mathematical culture that emerged after the Second World War.

The work of Evans will prove to be particularly interesting, for Evans's ideas were ignored not because they were mathematically unsophisticated. Nor was he disconnected from the disciplinary networks that validate acceptable contributions to the discipline. Rather, it is my contention that the marginalization of Evans's ideas in the postwar period in economics is better understood as a result of a change in the conception of what mathematics itself could bring to a scientific field. An implication of this reading is that any narrative in the history of economics of the twentieth century that employs the idea of "increasing mathematization" should be read with skepticism.

Vito Volterra's World[1]

In 1959, when Dover Press reprinted Vito Volterra's *The Theory of Functionals and of Integral and Integro-differential Equations*, Griffith Evans was asked to write the preface. In a lovely three-page note, he mentioned that "it was my good fortune to study under Professor Volterra from 1910 to 1912" (Evans 1959, 1). Evans was twenty-three at that time, and his 1910 Harvard Ph.D. earned him a Sheldon Traveling Fellowship, which he used for postdoctoral study at the University of Rome. This was to be the marker event in his intellectual life, for from his first published paper in 1909, "The Integral Equation of the Second Kind, of Volterra, with Singular Kernel," he was connected to the greatest of the Italian mathematicians of the Risorgimento, the intellectual leader of Italian science in the late nineteenth and first part of the twentieth century. Indeed, by 1911 Evans had published six papers in Italian, on functional analysis and integral equations, in Volterra's journal, *Rendiconto Accademia Lincei* (*The Proceedings of the Lincei Academy of Science, Physics, Mathematics, and Nature*). The interests developed in Volterra's Rome were to be the defining intellectual themes in his mathematical life, as his papers continued to explore both potential theory from a perspective of classical mechanics, and the theory of functionals. Along the way, Evans took time to write "The Physical Universe of Dante" in 1921, and to explain the Italian school of algebraic geometry to a large audience in "Enriques on Algebraic Geometry" in 1925. His last new paper, published in 1961 at age 74 (!), was "Funzioni armoniche polidrome ad infiniti valori nello spazio, con due curve di ramificazione di ordine uno." It is safe to say that his command of the Italian language, and his early connection to the mathematical subjects created by Volterra, linked Evans irrevocably to Volterra, who became his intellectual model. How else are we to read the gracious, and admiring passage:

> Volterra was close to the Risorgimento, close to its poets and their national ideals. In 1919, he was prevailed upon, in spite of a modest reluctance, to give a lecture on Carducci. I remember the occasion well because I had the pleasure of translating this, as well as his exposition of functions of composition, viva voce to an audience of students. He was also close to the Rinascento, with respect to his sensitivity to art and music and his unlimited scientific curiosity. His devotion to the history of science and his feeling for archeology were expressed respectively in the treasures of his personal library in Rome and in his collection of antiquities in his villa at Ariccia. He took a most prominent part in the international organizations of science and in extending the cultural relations of Italy. His career gives us confidence that the Renaissance ideal of a free and widely ranging knowledge will not vanish, however great the pressure of specialization. (Evans 1959, 3)

Volterra was the mathematician Evans wished to be; the mentoring of postdoctoral students is to this day frequently a process of professional modeling and career and interest shaping. In Evans's case, this was to be manifest in his life-long connection to Italy, and to Volterra himself.

We have the handwritten autobiographical notes that the aged and failing Evans tried to put together in 1967, fragments of an autobiography that he could not complete. He said that command of the Italian language was not gained early, or at home, for as a Harvard student of "Copy"— Charles Townsend Copeland—he frequently wrote literary themes and "could read and enjoy French, German and Latin as well as English" (Evans 1967). At another place in these disorganized notes, he recalls a series of images from the past, and in one place remarks that "towards the end of the war (Sept. 1918) I was up towards the front, on the Lido (if I remember correctly), to 'inspect' the Italian antiaircraft defenses. . . . Earlier somewhere near the mountains (Padova?) I remember however an open prairie, at the front way north of the Po (Battle of the Piave). Volterra and I were driven up in a big automobile to Padova and I was taken to the front" (ibid.). The point is clear. Even as memory and handwriting failed, and as his powers waned, Evans took great pride in the fact that his work and Volterra's were connected.

Since I shall argue that the interest that Evans took in the mathematization of economics can be informed by Vito Volterra's interest in the mathematization of economics, we need to know a bit more about Volterra.[2] He was a remarkable man. Born in 1860, he and his mother were left destitute when Volterra's father died in 1862. Relatives took them in, and Volterra grew up, and was educated, in Florence. He was very precocious, and his teachers quickly recognized his remarkable abilities. One of them, the physicist Antonio Roiti, intervened in the family discussions concerned with launching Volterra into a commercial career by making the high school boy his assistant in the physics laboratory at the University of Florence. From there, Vito Volterra won a competition to study mathematics and physics at the University of Pisa in 1880. Receiving his doctorate in 1882, with several publications in analysis in hand, he was appointed as assistant to the mathematician Betti, and the following year won a post as Professor of Mechanics at Pisa. In "1900 he succeeded Eugenio Beltrami in the chair of mathematical physics at the University of Rome" (Volterra 1976, 86). It was in Rome that he established his position as scientific leader-spokesperson of the new country. His research interests were wide-ranging, and his position in Rome allowed him to stay at the center of activity.

> Rome became the capital of the newly created state of Italy in 1870. Under the leadership of Quintano Sella (1827–1884), a mathematician at the University of Turin who exchanged academic life for a ministerial post in the new government, Rome's scientific halls came to life again. With the help of the new Commissioner for Public Instruction, also a mathematician, Sella brought the cream of Italy's scientific fac-

ulty to Rome and transformed the capital's historic Accademia dei Lincei into a genuine National Academy of Science. Sella and his colleagues built the scientific world that Vito Volterra, then 40, inherited in 1900 when he took up his duties . . . in the nation's capital. [Volterra's] appointment as Senator of the Kingdom five years later reinforced the Risorgimento tradition of the scientist-statesman in the service of king and country. (Goodstein 1984, 607–8)

Volterra was concerned with all aspects of scientific understanding. He was one of the leaders in making Einstein's theory known, and he was a tireless worker for the public appreciation of scientific knowledge. In our present day of "Science Wars" and of lost faith in the very notion of progress, it is difficult to recapture the optimistic world in which science, scientific knowledge, and technology or applied science was to lead to the new enlightenment. Volterra was the kind of new man to whom Henry Adams probably referred in "The Dynamo and the Virgin" (chapter 25 of Adams's autobiography), an instantiation of the ideal of the scientific cum technical polymath. His interests in mathematics, for example, did not prevent, at age fifty-five, his working as a lieutenant in the Italian Corps of Engineers, during which period "he worked at the Aeronautics Institute in Rome, carried out aerial warfare experiments on airships in Tuscany, and tested phototelemetric devices on the Austrian front" (Goodstein 1984, 610).

As a mathematician of the late nineteenth century, however, Volterra was of that generation so well described in the portrait of the fictitious Victor Jakob in *Night Thoughts of a Classical Physicist* (McCormmach 1982). Volterra was a classical analyst, whose mathematical work on functionals grew out of his generalization of differential equations to the more complex, and rich, theory of integro-differential equations: his mathematics papers in the period 1900–1913 include "Sur la stratification d'une masse fluide en equilibre" (1903), "Sur les equations differentielles du type parabolique" (1904), "Note on the application of the method of images to problems of vibrations" (1904), and "Sulle equazioni integro-differenziali" (1909).[3] Nevertheless, these topics were not so well connected to the nascent Italian school of algebraic geometry launched by Cremona, developed by Bertani, Segre, and Veronese, and brought to fruition by Castelnuovo, Enriques, and Severi. Volterra was thus more in the tradition of analysts like his mentor Betti, and Dini, for the former worked in the intersection of analysis and physics, and the latter was concerned with the rigorous reformulation of mathematical analysis. For Volterra, trained by both men, it was the case that

> his own research was more closely linked to the applications. In Volterra's view it is the peculiar problems deriving from the experimental sciences that lead to the most fertile and useful theories, while general questions posited in abstract terms often lack any applications. This conception of the relationships between analysis and physics are directly linked to the French physico-mathematical tradition from Fourier to Poincaré. In fact the need for concreteness apparent in the fact of thinking of mathematical issues as linked to physical problems, together with the rigorous training received in Dini's school, helped to make Volterra particularly well-suited to tackling mathematical physics. (Israel and Nurzia 1989, 114)

The physicist Victor Jakob of McCormmach's novel, though German, could represent the same kind of scientist in many of the European countries at that time. Relativity was a real shock to the traditional vision, and quanta were difficult to domesticate. The crises of atomic theory, and subatomic particles, were on the horizon, and the issue of what kind of theory would promise the best explanation brought into question the very idea of explanation itself. As Israel and Nurzia put the matter,[4]

> The crisis in question stemmed from the discussion going on in the scientific world of the time concerning the advisability of maintaining the classical mechanic method of explaining natural phenomena based on the deterministic principle, as well as on the mathematical tool provided by differential equations. From the strictly mathematical standpoint this crisis led to a split between "antiformalists," who favored a development of mathematics linked to experimental issues, and "formalists," who preferred development free from all constraints except formal rigor. The reactions to this crisis by the world of Italian mathematics varied enormously and a number of totally conflicting attitudes emerged. On the one hand there were those, like Volterra, who merely acknowledged the existence of a crisis and sought, if not a

solution to the crisis, at least a solid foundation in their scientific practice and in the links between mathematics and topics of experimental design. . . . [It is for this reason that] Volterra supported and promoted scientific organizations whose main purpose was to provide a concrete means of bridging the gap between pure and applied science. (Ibid., 115, 116)

Although Volterra appears not to have left autobiographical pieces, or much material that directly expresses his own particular views of the role of mathematics in applied sciences, we do have a document that, interpreted as a projection of his own views, may be helpful. This paper, delivered as an address at the inaugural festivities for Rice University (then the Rice Institute) appears to have gone unnoticed by those few economists who have become interested in Volterra. Delivered in French, the address was translated by Griffith Conrad Evans, and was titled simply "Henri Poincaré" (Volterra 1915). Poincaré had just died, and this address, Volterra's tribute to him, was both a eulogy and an appreciation for a scientific and personal career, a tribute that well suited the foundation of a new institute dedicated to science in America. In attempting to place Poincaré in history, in the history of mathematics and science, Volterra went in some detail to the history of the study of differential equations and function theory in the latter part of the nineteenth century:

There are two kinds of mathematical physics. Through ancient habit we regard them as belonging to a single branch and generally teach them in the same courses, but their natures are quite different. In most cases the people are greatly interested in one despise somewhat the other. The first kind consists in a difficult and subtle analysis connected with physical questions. Its scope is to solve in a complete and exact manner the problems which it presents to us. It endeavors also to demonstrate by rigorous methods statements which are fundamental for mathematical and logical points of view. I believe that I do not err when I say that many physicists look upon this mathematical flora as a collection of parasitic plants grown to the great tree of natural philosophy. . . . The other kind of mathematical physics has a less analytical character, but forms a subject inseparable from any consideration of phenomena. We could expect no progress in their study without the aid which this brings them. Could anybody imagine the electromagnetic theory of light, the experiments of Hertz and wireless telegraphy, without the mathematical analysis of Maxwell, which was responsible for their birth? Poincaré led in both kinds of mathematical physics. He was an extraordinary analyst, but he also had the mind of a physicist. (146–47)

What we have is Volterra's own projection onto Poincaré of the kinds of values that a mathematician ought to exhibit in his work: not just a mathematical sophistication and power of analytical reasoning, but a deep and thorough understanding of the scientific basis and connection of those mathematical ideas. Poincaré, mathematician and scientist, was Volterra's paradigmatic intellectual.

Volterra and Economic Theory

The distinction that Volterra makes in this passage between the two approaches to doing physics, the distinction between grounding explanations on the physical characteristics of the problem, or grounding explanations on mathematico-logico reasoning chains, mirrors the distinction between nonformalist and formalist responses within the mathematics community to the crisis of the foundations of mathematics, the paradoxes of set theory, of almost the exact same period. In the case of both physics and set theory mathematicians could, with the formalist response, ground the unknown upon the known. For mathematics, the grounding was to be an axiomatization of the settled parts of mathematics, logic, set theory, and arithmetic, as a basis of both more "advanced" mathematical theory and the sciences built upon the axiomatized mathematical structures so created. For Volterra, this formalist response was not rigorous: scientific reasoning chains had to be based not on the free play of ideas, or axioms, or abstract structures. Rather, scientific models had to be based directly and specifically on the underlying physical reality, a reality directly apprehended through experimentation and observation and thus interpersonally confirmable.

This point is important, and bears repeating because the present-day identification of rigor with axiomatics obscures the way the terms were being used at the turn of the twentieth century.[5] Today we tend to identify the abstract reasoning chains of formal mathematical work with the notion of rigor, and to set rigor off against informal reasoning chains. Unrigorous signifies, today, intellectual informality. This distinction was not alive in Volterra's nineteenth century world, however. For Volterra, to be rigorous in one's modeling of a phenomenon was to base the modeling directly and unambiguously on the experimental substrate of concrete results. The opposite of "rigorous" was not "informal" but rather "unconstrained." To provide a nonrigorous explanation or model in biology, or economics, or physics, or chemistry was to provide a model unconstrained by experimental data or by interpersonally confirmable observations.

Volterra sought to mathematize economics and biology by replacing metaphysical mathematical analogies with rigorous mathematical models.[6] In economics, however, Volterra has but one "official" publication, a review of Pareto: "L'economia matematica ed il nuovo manuale del prof. Pareto,"[7] (Volterra 1906a). For Vito Volterra, the strategy for approaching scientific explanation generally was to base reasoning on the most well-developed intellectual framework then extant, the framework of classical mathematical physics. His clearest statement of this position, of special interest to economists, is the 1901 paper "Sui tentativi di applicazione della Mathematiche alle scienze biologiche e sociale."[8] It will be useful to examine one lengthy passage from this paper, for in it Volterra defines what, at the turn of the century, the position entailed for the field of economics. Toward the end of that piece, Volterra cautioned:

> The notion of homo oeconomicus which has given rise to much debate and has created so many difficulties, and which some people are still loth [sic] to accept, appears so easy to our mechanical scientist that he is taken aback at other people's surprise at this ideal, schematic being. He sees the concept of homo oeconomicus as analogous to those which are so familiar to him as a result of long habitual use. He is accustomed to idealizing surfaces, considering them to be frictionless, accepting lines to be nonextendable and solid bodies to be nondeformable, and he is used to replacing natural fluids with perfect liquids and gases. Not only is this second nature to him: he also knows

the advantages that derive from these concepts. If the mechanics scholar pursues this study he will see that both in his own science and in economics everything can be reduced to an interplay of trends and constraints—the latter restricting the former which react by generating tensions. It is from this interplay that equilibrium or movement stems, one static and one dynamic, in both these sciences. We have already referred to the vicissitudes of the idea of force in the history of mechanics: from the peaks of metaphysics we have descended to the sphere of measurable things. In economics, for example, we no longer speak as Jevons did about the mathematical expression of non-measurable quantities. Even Pareto seems to have given up his idea of ophelimity, which was the cornerstone of his original edifice, and is moving to purely quantitative concepts with indifference curves which so beautifully match the level curves and equipotential surfaces of mechanics. . . . Lastly our mechanical scientist sees in the logical process for obtaining the conditions for economic equilibrium the same reasoning he himself uses to establish the principle of virtual work, and when he comes across the economic differential equations he feels the urge to apply to them the integration methods which he knows work so well. (Volterra 1906b, 9–10)

By rigorously solving well-defined problems in a clearly delimited field, mathematical economics must offer us a secure foundation of positive data on which to base our judgement as to the procedures to be followed in various circumstances. But it always leaves open the discussion of the great moral and political questions to which such results should be applied. . . . But to ensure that one can fully justify the application of mathematics and to obtain the secure results one seeks, it is first of all necessary that the problems be formulated clearly and [be] based on definitions and postulates containing nothing vague. It is also essential that . . . the elements taken into consideration are treated as quantities that cannot elude measurement. (Volterra 1957, 142, 144)

In Israel's view (Israel 1988, 1991a, 1991b; Israel and Nurzia 1989), Volterra's view of science and scientific explanation, which entailed rigor in modeling in the sense of developing economic explanations from mechanical ones, came up against the nonempirical nature of economics, and the impossibility of erecting mathematical economic theories on any empirical foundation whatsoever.

The larger issue, however, was that Volterra's perspective was increasingly unsatisfactory as a solution to the crisis in the natural sciences. Indeed the entire crisis, at least in physics, turned on the explanatory power of mechanical reductionism. Far from being part of the solution, reductionist thinking such as Volterra's was itself the problem. The crisis, or rather the interlocked crises, of mathematics and physics was resolved by the formalist position on explanation whereby mathematical analogy replaced mechanical analogy, and mathematical models were cut loose from their physical underpinnings in mechanics. The result was that in the first decades of the twentieth century a rigorous argument was reconceptualized as a logically consistent argument instead of as an argument that connected the problematic phenomenon to a physical phenomenon by use of empirical data: propositions were henceforth to be "true" within the system considered (because they were consistent with the assumptions) and not "true" because they could be grounded in "real phenomena." We can leave Volterra here, and refocus on Evans, for a historian of economics can construct Evans out of these Volterra-emergent themes.

Evans: The Mathematician and His Interests

In 1912 Evans became one of the first two teachers at the Rice Institute in Houston. As Rice was transformed into Rice University, Evans lent his increasing renown and intellectual strength, and mathematical visibility, to the new institution. Today Rice recognizes his role in its program of Griffith C. Evans Instructorships in Mathematics intended for promising young mathematicians. Evans's resignation from Rice in 1933 was sufficiently noteworthy that the *Houston Chronicle* wrote a news story recounting his career there, for he had, over his twenty-one years at Rice and in Houston, made his mark. After his promotion to full professor in 1916, "he was married in 1917 to Isabel Mary John, daughter of state court judge and Mrs. Robert A. John of Houston. Judge John was General Counsel of the Texas Company for many years" (*Houston Chronicle* 1973). Isabel John was a great-granddaughter of Sam Houston, and her niece became Mrs. Price Daniel, whose husband was a Texas governor; Griffith Evans, Boston Brahmin, was nothing if not well-connected to the first families of Texas.

The Rice University Archives provide a glimpse of Evans's diverse tastes and interests. On 6 May 1915, the *Houston Chronicle* reported on a lecture that Evans gave on Pragmatism "which was the first of a series of three lectures on 'scientific aspects of philosophy.'" Evans began by addressing the gospel of Tolstoy, and then suggested that "in regard to such basic [metaphysical] questions almost all thinkers have a basis of optimism. Their query is not 'are things right?' but 'how is it that all things are right?' and in their researches they seem to trust to what may be called the lucky star of humanity, injecting their personal interest in the outcome into the problem itself." Evans went on to discuss William James and his approach to settling or at least posing philosophical issues. He concluded that "we have no reason to suppose that all possible phenomena can be expressed by means of any finite system of terms. Instead of this, we may expect that, no matter how complete our system of conceptual terms may be, we shall find facts that require its continual extension. That is what we mean when we say that there will always be novelty in the world, and always new problems for the genius of man to attack and solve" (*Houston Chronicle* 1915a).

One week later, the *Houston Chronicle* of 13 May 1915 reported on Evans's talk on Aesthetics, the second in that series of three lectures. Evans began by suggesting that there was an important element in a discussion of art and aesthetics that could be approached through understanding how wide the aesthetic net should be cast. He suggested that knowledge of the aesthetic issues in mathematics could cast light on the general problem. The inability of mathematics itself to determine which geometry is "correct" leads to a position "that an arbitrary element enters into that most exact of sciences, mathematics. . . . The nature of mathematics is that it is entirely arbitrary, and its use is that in its growth, by the formation of arbitrary concepts, it limits itself more or less unconsciously to those who have mirrors in actual life. It is therefore a 'human interest' story."

Beyond the technical skills required to be a mathematician, "other qualities of a far more subtle sort, chief among them which is imagination, are necessary." He cites Benedetto Croce with approval and remarks, "Art is expressed in intuition, that is, the synthesis of concrete imaginative elements. It is a spiritual, theoretical activity."

The third Evans lecture on "Scientific Aspects of Philosophy" was reported in the *Houston Chronicle* of 28 January 1916. This final lecture, on Rationalism, began by noting, "Each of the earlier two lectures ended with a problem which could not be solved in terms of the methods proper to subject of the lecture itself." In the third lecture, Evans framed the issues as fundamentally epistemological: paraphrasing Kant, he asked, "How is metaphysics possible? How can we hope to know anything about metaphysical questions?" Kant further asked whether natural science is possible in the sense that natural laws are not law like of experience, but are rather creations of the human mind. Evans paraphrases Kant with approval and remarks that

> the axioms of mathematics are merely the forms in which sensations must be presented to us in order to become mental representations, they are meshes through which our intuitions are formed, and derive their necessity from that fact. . . . Similarly the laws of the pure science of nature are merely the laws of the understanding. By means of them nature becomes intelligible. And they derive their certainty from that reason.

These three lectures, never to my knowledge printed in full, aid our understanding of Evans's perspective on the role of mathematics as a human activity, and thus as an activity entirely appropriate to be connected to other human activities like Marshall's "study of mankind in the ordinary business of life." Evans is an end-of-the-nineteenth-century rationalist, a Harvard pragmatist who believes in reason with a human face, and man's capacity to understand the world in which he lives. For many mathematicians and physicists, the earth had moved in the 1890s: Evans was writing on the eve of World War I, a time in which civilization, and its products, would be shaken. There is no trace in Evans's own work of the intellectual crisis that so rocked the turn-of-the century physicists and mathematicians: Evans's scientific views remained intact.

Evans's war career was recorded in the Houston newspapers (1918), which while noting his appointment as "scientific attaché to the American Embassy in Rome, Italy," observed that Captain Evans was doing "research with the American Aviation Service. He was in Italy with the American and allied forces studying actual conditions. . . . He has been in war work in France, England, and Italy since March, his ability as a linguist adding to his proficiency as a scientist in foreign countries." The paper followed this story with another (1919), on Harvard's offer of a faculty post to Evans, which, upon his demobilization, he turned down to return to Houston.

We can reconstruct Evans's career and interests on his return to Houston with three articles he prepared for *The Rice Institute Pamphlet*. The first of these, in April 1921, appeared in a series of seven lay lectures observing the six-hundredth anniversary of the death of Dante to which Evans contributed a substantial paper on "The Physical Universe of Dante" (Evans 1921). For me this paper best places Evans as an unusual scholar, though he acknowledges the help of Vito Volterra and other individuals from Rome as well as his own father and Professor Tyler of Boston for assistance. In the paper, Evans examines the context in which Dante wrote, situating Dante in a particular intellectual framework of knowledge about the physical universe. With a sharp command of the original source material, as well as wide-ranging knowledge of the secondary literatures in the history of science, Evans discusses the issues of the calendar, and astronomy, to locate allusions and references in Dante's poetry, and even delves into astrology as an interpretive system for comprehending the nature of Dante's physical universe. Evans continues his discussion of science with observations on biology and physics and their role, together with some observations about the "discovery" of petroleum! He discusses, for example, "a striking error in Dante's notion of the civilized world is that of making the length of the Mediterranean extend for 90 degrees of longitude—perhaps an intentional remodeling of geography to fit allegorical interpretations." This is connected to Evans's subsequent discussion of the geometrical properties of the earth's surface: "witness the geodesy of the Third Tractate of the 'Convivio,' where the relative positions of poles, equator, and elliptic are discussed, and the relation of day to night. Here incidentally the radius of the earth is given as 3250 miles." Evans then continues with discussions of the relation of Dante's references to the heavenly motions to the then popular theories of the variable apparent motions of the heavens created "out of a system of uniform circular motions about centers themselves also moving uniformly." Evans concludes the lecture by building up a comprehensive model of the universe, the motion of the stars, the sun, the planets, and the moon from the medieval conceptions and as present in Dante's own discussion. However, in his last two paragraphs Evans takes a delicious Whiggish turn:

It is time perhaps for science to grow beyond the need of a mechanical interpretation. . . . Whenever there is one mechanical explanation, the transformation theory of dynamics tells us that there is more than one, and of these the simplest as Einstein has shown us, is the most complicated. On the other hand, when we try to classify the phenomena that admit of mechanical explanation, and Professor G. D. Birkhoff tells us that any system of ordinary differential equations is nothing but a set of dynamical equations, and vice versa, it becomes evident that the future of science may soar farther from our own restricted mechanical point-of-view than ours has risen above the quaint interpretations of the middle ages.

This 1921 pamphlet followed an October 1920 paper by Evans in the same pamphlet series: "Fundamental Points of Potential Theory." The occasion for Evans's memoir on the subject were "three lectures delivered at the Rice Institute in the Autumn of 1919, by Senator Vito Volterra, Professor of Mathematical Physics and Celestial Mechanics, and Dean of the Faculty of Sciences of the University of Rome."

Evans's own paper is "a study of the Stieltjes integral in connection with potential theory." In the paper he demonstrates the relation of the potential function thus defined to the integral form of Poisson's equation, which applies to any sort of distribution of mass. Evans's object is the set of general forms of Green's theorem as applied to polarization vectors and solutions of Poisson's equation. The paper concludes with a study of the appropriate boundary value problems for harmonic functions and the general open region. The investigations represent, as he notes, "studies originated in 1907, when it first became apparent to me that the theory was unnecessarily complicated by the form of the Laplacian operator."

The third article provided by Evans to the *Rice Institute Pamphlet* appeared in volume 13, no. 1 of January 1926. In this series of five lectures observing the three-hundredth anniversary of the death of Francis Bacon, Evans contributed a paper on "the place of Francis Bacon in the history of scientific method." This paper on Bacon is an interesting one, as it captures Evans's own philosophical ambivalence. Evans, as a resident intellectual cum philosopher/historian of science, although an amateur, seemed to be obliged to contribute to a set of popular lectures on Francis Bacon. However, while the mathematician in Evans really had no patience with Bacon and thought he was not worth taking seriously, the respectful scholar in Evans needed to take Bacon seriously, as he recognized the esteem in which Bacon was held by such figures as Liebnitz, Locke, Hume, and Kant. Hence the paper's implicit dilemma.

Evans "solves" this problem by walking away from it. He begins with several pages attempting to answer the question, "If there had been no Bacon, would the future of science been essentially different, or would its development have been materially slower?" He gives the game away by immediately answering this question himself with "I think we may give the negative answer to both these questions." He then proceeds to examine several of the researches that could have claimed Bacon in a line of paternity, as it were. In none of them does Bacon really play a role. Neither physics, nor astronomy, nor Atomistic or relativistic theories, nor theories of electricity and magnetism, certainly not mechanics, not any of these fields, in Evans's terms, "pass close to Bacon" for "given Bacon's neglect of mathematics, it is not surprising that these mathematical methods go back on a line which Bacon does not grasp." In addition, certainly in neither evolution nor biology, nor in chemical investigations, did Bacon take much scientific part.

It is not for the science that Bacon deserves to be recalled. Rather, Evans locates him in a line with the great "skeptic Montaigne." For this French philosopher, "reason is to him a dangerous tool, and he who uses it loses himself along with his dogmatic enemies." Evans locates his subject:

> Francis Bacon believes that he provides the way of putting in order the universe which Montaigne has left in such an unhappy state. He devises a method which he thinks will be easy to apply and will increase the domain of science enormously and rapidly. . . . Bacon tends to diminish the importance of the imagination in arriving at scientific truth. . . . What is to be the real method of turning natural history into science is the systematic use of reason in the way in which Bacon explains . . . as an induction with the help of experiment. According to Bacon's idea it is possible to arrive at a scientific theory by a process of exclusion, more or less as an argument by reductio ad absurdum is used in mathematics. . . . In other words, hypotheses are to be eliminated successively with reference to fact or experiment until only the hypothesis which must be true remains.

Evans goes on a bit with this discussion but returns to the point that clearly gnaws at him, that there is no place for the imagination. "Knowledge is to be advanced by the invention of new concepts. But what makes a concept significant?" Evans here takes his stand with Liebnitz and in what is certainly his own voice remarks, "It is brilliance of imagination which makes the glory of science." However, if Bacon had a deficient imagination, perhaps Dante had too much for the kinds of rigorous connections via creative hypotheses to the empirical world that a scientist, such as Evans, thought he should make.

Evans among the Economists

At a first pass, we can locate Griffith Evans's connection to economics through the sequence of papers he wrote that led up to his 1930 book, *Mathematical Introduction to Economics*. These five papers all appeared in regular mathematics publications, and all essentially operated in the same fashion: they called the attention of mathematicians to interesting problems in an applied discipline.

The first of these papers was entitled "A Simple Theory of Competition" and appeared in the *American Mathematical Monthly* in 1922. In it, Evans postulated a rudimentary theory of competition in terms of specific functional forms. Basing his discussion on Cournot's volume, and developing the discussion in terms of the profits of several competitors, Evans handles a number of special cases. From a modern point of view, the interesting feature of Evans's discussion is that he works with quadratic cost functions and something akin to a linear demand function. His analysis operates entirely independent of a decision calculus for either producers or purchasers. With the three coefficients of the cost functions, and the two coefficients for the demand function, there is a variety of special cases that can emerge under different assumptions. Evans modifies his discussion by introducing more producers, different kinds of taxes, and other specifications of the cost curve. He ends by noting that the restriction to functions of a single variable is mathematically inessential. The deeper question is, "What is retained when we remember that what a producer is interested in is not to make his momentary profit a maximum, but his total profit over a period of time, of considerable extent, with reference to cost functions which are themselves changing as a whole with respect to time?" (379) Evans notes that this leads to problems in the calculus of variations and concludes the paper with his "regrets that at the present time he can refer only to his lecture courses for a further treatment of this point-of-view. Nevertheless it seems the most fruitful way that a really theoretical economics may be developed" (380).

The next paper in this sequence appeared in 1924, again in the *American Mathematical Monthly*. "The Dynamics of Monopoly" picked up the theme developed in the 1922 paper of change over time. Evans assumes, for a monopolist, an interest in making total profits as large as possible over a time interval. With an initial price, and a final price, Evans sets up the problem of maximizing the appropriate integral. Following the statement of the problem, which refers back to Amoroso's 1921 discussion of economic dynamics, Evans states that "an editor of the Monthly—Professor Bennett—has said that one should be obliged to present a certificate of character before being initiated into the mysteries of the calculus of variations, to which study our present investigation belongs, since its fascination is so great that neophytes seek to introduce it into problems which would otherwise be perfectly simple."

Evans then proceeds to make the matter of the dynamic behavior a sequence of discussions of special cases. He develops what he calls the Cournot monopoly price as one kind of solution associated with an appropriate end value. Most interesting, however, to a modern reader, is his concluding section, which notes that "one purpose in writing the present paper, as well as the previous one, has been to show the wide range of problems suggested are solvable by a moderate mathematical equipment, and to encourage others to read in a direction that cannot but be fruitful." That sentence has a footnote, which reads, "For example, the works of Cournot, Jevons, Walras, Pareto, and Fisher. Those who can read Italian will find interesting the volume of Amaroso, already cited" (83).

One begins to see in this discussion Evans, the applied mathematician, finding problems to solve in the field of economics, some of which having already been treated by what, to a mathematician, would be primitive mathematical techniques. For Evans, behavioral rules and theories do not appear, nor does there appear a theory of price formation in markets. Rather, there emerges a discussion of output levels associated with different interrelationships among producers, under a variety of cost curve assumptions.

Evans's first 1925 paper, "Economics and the Calculus of Variations," appeared in *The Proceedings of the National Academy of Sciences*. It is a very different kind of paper in that it presents a theory of the interconnection between economic modeling procedures and the calculus of variations as a mathematical structure. Evans here operates at a quite different level of generality from the papers on competition and monopoly, presenting a general systems vision.

> The writer is not the first to venture to state a general theory in mathematical terms of a subject which is not unfairly regarded as compounded somewhat indefinitely of psychology, ethics, and chance. Being more than a mere mixture, however, it is equally fair to say that a separate analysis may apply; indeed, in Economics we are interested in the body of laws or deductions which may be inferred from convenient or arbitrary economic hypothesis, however they may be founded—in fact, fiction, statistics, habits or morals—what we will. This process of inference, if it is worthy of the effort, may be made mathematical.

Evans develops the notion of an abstract economy, by dividing an economic system into a set of n compartments, and letting dx_i/dt be "the rate at which the specific commodity or service i is produced in its compartment." Defining a rate at which this commodity comes out from and comes into the compartment and a rate at which it is present within the compartment and noting that there is a balance among these three rates, and that there is an input-output accounting identity at work, Evans defines a general system of economics as a set of M laws linking the behavior of the flow variables over time. Evans develops the flows and the balances over time in the framework of the calculus of variations, examining money, and the equation of exchange, in this context.

He concludes, "It may be remarked that the relation of economics to the calculus of variations is not accidental, nor the result of the generalization from previously found differential equations, since it is in the nature of an economic system that there should be a striving for a maximum of some sort." In this paper, Evans remarks in a footnote that "Mr. C. F. Roos, in an article not yet published, treats a . . . problem of a similar nature." Roos was, of course, Evans's student at Rice, and one of the founding members of the Econometric Society. This paper on the calculus of variations and economics solves not a particular problem but rather frames a conceptual one: how one might model an economy or economic system. What Evans accomplishes here, though it stands outside usual schemes, is a dynamic input-output model of an economic system. Moreover, this dynamic model is mathematically coherent, and rich enough to permit some inferences.

Evans's remarkable "The Mathematical Theory of Economics" paper, which appeared in the *American Mathematical Monthly,* also in 1925, was originally read at the December annual meeting of the American Mathematical Association in Washington, D.C., in 1924.

> One interest in research in the Mathematical Theory of Economics is that the necessary preparation for it either in mathematics or in economics is not so great as for theoretical research say in physics or chemistry, or even in biology. . . . It may well be that it is the lack of mathematical technique among economists which has prevented the theoretical side of the subject from developing as rapidly as the wealth of books and papers, devoted to it, would seem to indicate. On the other hand, if we turn to the trained mathematicians, we find them mainly engrossed in the more romantic fields of physics, chemistry, and engineering, except in the case of the extensive analysis of statistics, where contact is made with kinetic theory on the one hand, and social and biological data on the other. (104–5)

In the next several sections, Evans lays out Cournot's views on monopoly and competition, and those of Irving Fisher's, more or less restating the results of his earlier papers in the *American Mathematical Monthly.* Additionally, he refers to the calculus of variations argument, and notes Roos's paper, which will shortly appear in the *American Journal of Mathematics* (Roos 1925). Of interest, however, is the remarkable concluding section five, "General Points of View." Here we begin to see why Evans was to be so much an outlier among economists. He writes:

> There is no such measurable quantity as "value" or "utility" (with all due respect to Jevons, Walras, and others) and there is no evaluation of "the greatest happiness for the greatest number"; or more flatly,—there is no such thing. In a way, material happiness has to do with a maximum of production and a minimum of unpleasant labor; though again no such thing is realizable theoretically without an arbitrary

definition of a composite function which is to take on a maximum value; and in the composition of this function the labor and profit of various classes of people enter capriciously. One might define a ratio of weighted production divided by weighted amounts of labor, according to classes, and study what sort of lash, economic or otherwise, would serve to impel Society towards this limit; but the choice of weights would depend essentially on whether the chooser is born a Bolshevik or a member of the Grand Old Party! Compromises carry us into the field of ethics.

That does not mean that such study is unprofitable. Far from it. How otherwise are we to evaluate the schemes of reformers and prophets, major and minor? Moreover the groundwork of such studies must be made well in advance, before there is any direct occasion for them; otherwise they will fail us when we do need them. There is not only an opportunity for mathematics and economics, but even a duty; and on mathematicians in an unusual degree lies the responsibility for the economic welfare of the world. (110)

Thus certainly by 1925, five years before Evans's book on mathematics for economics, he has written himself outside the usual concerns of later (post-World War II) economists. He is dismissive, if not virtually contemptuous, of the intellectual framework upon which neoclassical analysis had been founded: the subjective theory of value. For Evans economists, even mathematical economists like Jevons, Walras, and most certainly Marshall, were on the wrong track and had little useful to contribute if they believed in the analysis of value or utility. Evans here takes on the crudest of materialist positions, choosing to operate his analysis strictly in terms of production and labor quantities because for him, as for Volterra, these ideas could be linked to measurable quantities. For an anti-Marxist patrician of an old Boston family, this position is interesting indeed.

Just in case economists did not get the point, Evans, writing in 1929 in *The Bulletin of the American Mathematical Society,* reviewed the 1927 Macmillan edition of Cournot's book. Beginning his review by citing Marshall and Mill on Cournot's genius, he proceeds to distinguish Cournot's great understanding and insight from more recent treatments in mathematical economics. He does this by contrasting a "Cournot [who] is almost alone in holding to a clear realization of the difference between measurable and nonmeasurable quantities" with more recent authors who do not. Evans writes that "one recent book on the mathematical principles of economics, typical of many others, builds its theory on the following basis: Write $U(x,y, \ldots)$ for an algebraic function of measurable quantities." That author's utility discussion is linked to changes in utility and therefore satisfaction. "Apparently this other author is unaware that he is begging the question. If loci of indifference are expressed by Pfaffian differential equations it does not follow that there is any function of which these are the level loci for such equations are not necessarily completely integrable. The question is not of names, but of existence. These supposedly general treatments are much more special than their authors imagined." Evans's contempt for the misguided mathematically economist is quite open: he identifies the economist who said those silly things as "Bowley, *Mathematical Groundwork of Economics*, Oxford, 1924, p. 1."

Thus by the time Evans releases his book, *Mathematical Introduction to Economics*, into the world of economists in 1930, he is on record in print as believing that Jevons and his school, which of course means Marshall, and Walras and his school, which of course means Pareto, H. L. Moore, and virtually everyone else writing in mathematical terms, are entirely misguided for basing analysis on a non-quantifiable theory of value. Moreover, Evans has sneered in public at the mathematical competence of the individual, in England, who had written the basic text in mathematics for economics.

Mathematical Introduction to Economics

It is, of course, primarily for his book *Mathematical Introduction to Economics* (1930) that economists remember Griffith Conrad Evans. Written while he was still at Rice University, before his move to Berkeley, Evans's book represents an unusual conglomeration of topics and perspectives. The first several chapters, on monopoly, units of measurement, competition, price, cost and demand and taxation, take up themes and specific examples that Evans had introduced in his earlier publications on approaches of mathematics to economic problems. These chapters all have the Evans "hand" on them in their use of specific functional forms, and in their deliberate avoidance of behavioral assumptions and statistical work. Instead, these analyses reflect an interest with specifying the market outcomes, and developing relationships among variables to generate realistic or comprehensible special cases. Chapters 6 through 9, on tariffs, rent, rates of exchange, the theory of interest, and the equation of exchange and price level indices, operate at a slightly higher level of systematic abstraction, though for Evans, the treatment of these issues proceeds exactly as does the case of sales in a particular market.

It is, though, in chapters 10 through 12 that Evans takes his stand against the usual argumentation of economic theory. In chapter 10, for example, he states that "we must adopt a cautious attitude toward comprehensive theories." Arguing that while it is a temptation "to generalize a particular set of relations which has been found useful, by substituting variables for all the constants in the equation. . . . It may be questioned as to whether we have added to anything but our mathematical difficulties." It is not that he wishes to "abandon the search for general theories" but rather that "we shall gain much if we can formulate our propositions in such a way as to make evident the limitations of the theory itself." In short, "our endeavor then should be to make systematic discussions of several groups of economic situations, as theoretical investigations, and bring out the respective hypothesis which separate these groups" (110–11).

Evans uses this general discussion as a prelude to chapter 11's attack on economists (Jevons, Pareto, Walras) in their use of utility theory. He argues that those "authors with whom we are concerned . . . affirm that the use of mathematics need not be confined to [actual quantities of commodities and money] but may also be applied to the order relations among the subjective quantities." Those subjective quantities involve pleasure, satisfactions, and vanities. Evans refocuses his attack on utility theory through the integrability problem, the impossibility of building indifference surfaces from local optimization solutions.

In what must probably be a core set of intellectual principles which guided Evans in his own thinking, he remarks in subsection 67 (121–22) that

> a mathematical critique similar to that just adopted is widely explicable, and is more penetrating than an analysis in terms of loose concepts where the words themselves, by their connotations, may apply theorems of existence which are untenable . . . the concepts of beauty,

truth and good are analogous to those which we have been discussing. In every situation, there is something not of the best—some ugliness, some falsity or some evil—and so the practical judgement which is to be a basis of action is not "what situation is absolutely correct?" but "Which of several situations is best?" The problem involved is the comparison of two or more groups of elements of esthetic character. By the possibility of making a judgement at all is implied the fact that between two such groups, which are not too widely separated or which are simple in the sense of containing few enough elements, one can assign greater value to the one than to the other.

Evans takes the integrability argument, the most important topic in "formal" economics in the 1920s and 1930s, to mean that

we can devise an approximate value function as a scale for small changes of the variables, but cannot extend it beyond a merely local field unless we are willing to make some transcendental hypothesis about the existence of such a function. . . . In experimental terms we are accordingly not permitted to use such terms as beauty, good and truth with any absolute significance; comparative adjectives would better, or truer and these only as applied to situations which did not differ widely or differed only in one or two elements.

Not for Evans the grand unifying theory in aesthetics of George D. Birkhoff's *Aesthetic Measure,* or the unification of value theory of Edgeworth's *Mathematical Psychics.* Evans the mathematician, interested in potential functions and integral equations, is rooted clearly and distinctly in the physical phenomenon of measurable entities. If one can build a theory out of these bricks, well and good. However, if the mathematics precludes the building, one must not rush ahead and assume the building is already there.

Evans Marginalized

The Evans papers contain some materials related to Evans's view of his book and some attempts to manage its reception.[9] There is, for example, an undated handwritten letter probably to his publisher or editor H. J. Kelly at

McGraw-Hill about his desire to have Professor Roos review the book for the *Bulletin of the American Mathematical Society*. He also suggests that Professor Snyder of Yale thinks the book can be used "in connection with their mathematical club. Professor Kellogg thinks he can use it in connection with tutorial work. In general the use as a text must come slowly, since such courses are just beginning in the universities—Cornell and Yale are the only ones (besides Rice, where I have four students)."

> I have not read Hotelling's review. It would probably irritate me, if he did not see what the book is for. . . . It is the only book in the subject with exercises which the student can practice on, and the only book in English which consistently keeps to a uniform level of mathematical preparation using mathematics correctly. . . . The level of training is that which the engineer possesses. . . . It seems to me that you rule out your most important clientele, namely the large number of engineers who usually buy your books.

The papers also contain two letters from Henry Schultz at Chicago, of 24 April 1931 and 8 May 1931 the first of which replies to an Evans letter of 20 April. It would appear that the exchange developed out of Evans's learning that Schultz was to write a review, and seems to be based on a draft of that review sent to Evans. It also would appear that Schultz was concerned that the level of mathematical sophistication Evans assumed would be too high for most students of economics though he does express admiration for Evans's treatment of a number of topics, in particular the dynamic problems approached through the calculus of variations. In a most interesting remark, Schultz in his first letter says, "Frankly, I am puzzled by your attitude. In my naivete, I assumed that Volterra and Pareto had reached an understanding on this question"; after giving the Italian reference to the 1906 exchange, Schultz asks "Am I wrong? Are Pareto's revised views on utility and indifference curves—a revision which was necessitated by Volterra's criticism—still open to objection? If so, what is it? I should greatly appreciate further light on this question." He concludes this letter with "I am awaiting your reply to my query regarding Pareto's mature views on utility." Evans's reply is not to be found in the papers, but Schultz's letter of 8 May 1931 begins "I am glad to get your letter of May 2 and to find that we are beginning to understand each other." Schultz refers to a story that Evans must have told in the 2 May letter about his "experience with a chemical firm." Schultz, in counterpoint, describes his own experience with data and fitting curves to data and concludes, "It appears that any attempt to get light on coefficients of production or business methods is at this stage likely to be unsuccessful."

It would thus appear that the gist of Evans's views of utility involve issues of measures of utility or value. The nonquantifiable, the nonmeasurable, were hardly fit subjects for mathematical investigations from Evans's point of view. To one trained in *his* manner in mathematics, a mathematical theory of value and utility would necessarily be non-rigorous. Rigor, recall, is associated with the connection of the conceptual categories in an underlying physical reality. Rigor most decidedly did not mean for Evans what it meant for later mathematicians, namely "derivable from an axiomatization in a formal or formally consistent manner. The mathematician and mathematical economist Evans of 1930 is thus well connected to the Evans who studied with Vito Volterra before the First World War, and to the Volterra who abandoned economics in its non-rigorous infancy just after the turn of the century.

Despite Evans's marginal position within the community of economists, it should be noted that his papers confirm his own participation in the nascent subcommunity of mathematical economists. Charles Roos, at Cornell in the 1930s, had been Evans's student at Rice. When Fisher, Frisch, and Roos wrote to solicit organizational support for the creation of the Econometrics Society, Evans replied almost immediately with his support, and with the suggestions of individuals outside the United States to write including Schams of Vienna, Tinbergen, Leontief, and Rosenstein-Rodan. He was reading the works of economists who used mathematics whether they wrote in French, Italian, or German, and his reading lists suggest a broad-ranging intellect interested in keeping his courses up-to-date and his students well informed.

Given Griffith Evans's views on the state of mathematical economics, and the basis on which neoclassical theory had been constructed, the early reviews of the book were predictable. Writing in *Economica* (old series) in February 1931, R. G. D. Allen concluded that "the book contains many instructive applications of mathematics to economic problems, but, as a whole, it is not a convenient introduction to mathematical economics either for the pure mathematician or for the economist. The latter will be deterred by the lengthy algebraic development and, in the later chapters, by the difficult mathematical analysis used; the former, after a general survey of the work of Cournot, Jevons, and Walras, will be well advised to proceed, at once, to the complete analysis of Pareto."

In addition, in March of the following year, in volume 42 of the *Economic Journal,* the reviewer notes

> This book is interesting as showing a mathematician's approach to economics . . . but since there is no clear thread of economic theory in the treatment and no attempt at a general theory of any wide region of economics, a mathematician without economic knowledge will not obtain any thorough grasp of that subject; while the trained economic student will find the mathematical treatment difficult and in many places of a quite advanced level, while he will be bothered by the unelucidated mathematical character of the solutions. In fact, the appendix to Marshall's *Principles of Economics* is far more useful to the student of economics, quite apart from more recent studies on mathematical economics.

That reviewer was A. L. Bowley, and how could Evans really expect otherwise?

Evans was not easily put off. In what remains one of the most interesting, and prescient, critiques of the foundation of neoclassical theory, Evans took up "the role of hypothesis in economic theory" in *Science* on 25 March 1932. This paper had previously been delivered at a joint session of the Econometrics Society and Section K of the American Association for the Advancement Science on January 1, 1932.

The neutrality and generality of the title of Evans's paper belies its subversive intent. He begins by making a distinction between a natural and a theoretical science, the difference "lies essentially in the presence or absence of a free spirit of making hypotheses and definitions." In a theoretical science, as opposed to a natural one, definitions "become constructive rather than denotative and hypotheses are introduced and tried out, in order to see what sort of results may be reduced from them" (321). Evans then goes on to phrase the question as to "the degree to which we may speak of a theoretical economics, and the extent to which we may call it mathematical" (ibid.).

Evans develops his argument by suggesting that "the main object of economic theory is to make hypotheses, to see what relations and deductions follow from such hypotheses, and finally, by testing the consequences in comparison with the facts of existing economic systems, to describe them in terms of those hypotheses" (322). His illustration is demand as a concept. He presents five separate demand functions, each of which embeds specific assumptions or hypotheses. For example, one demand function might have quantity as a function of price alone, while a second would have quantity depending on both price and the rate of chance of price. In modern parlance, Evans is suggesting that we have a great deal of freedom in specifying the demand function.

The main line of argument follows quickly. "A simple concept in economics has been that of utility . . . but underneath such a definition there must lie assumptions, tacit or explicit. Even though we are not willing to assume that this psychic quantity is directly measurable, if we are to use it in equations we have nevertheless to be able to add small increments of it" (322).

He proceeds to suggest that in standard analysis "we leave out of account the question as to whether or not utility is itself measurable, but suppose that there is a quantity associated with it which is measurable and whose measure we may call an index of utility" (323). Evans continues by suggesting that a situation, described by a vector x_1, y_1, z_1 is not compared directly with a second situation x_2, y_2, z_2 but that rather if I is the utility index of one state, we examine dI as decomposed into the x, y, and z changes, as equation $dI = Xdx + Ydy + Zdz$. Evans argues that this is the actual comparison problem. Consequently, one must recapture the index of utility, I, from this equation: "In other words, we can build up an index function by means of the curves of indifference. But if the state of the system is given by three or more numbers, we also know that there does not exist in general such an index function. The expression of this fact in mathematical terms is the statement that an equation like $Xdx + Ydy + Zdz = 0$ is not completely integrable. If we wish to have a utility function, we must introduce some hypothesis on the coefficients X, Y, Z."

The problem is that, mathematically, such a process requires "that certain relations already hold between the variables x, y, z; and they are no longer independent. . . . Hence we must assume that all our situations relative to a utility function must not contain more than two independent variables, or else we must introduce directly a postulate of integrability. It seems an arbitrary limitation" (323–24). This, of course, is exactly the argument that Volterra (1906a) made to Pareto (Hands and Mirowski 1998).

Evans notes that economists have sometimes argued that there are, in fact, sufficient relations within the system among the variables to solve this problem. Sufficient in this case means that, for economists, there are as many equations as there are unknowns. However, "it is absolutely no check on the correctness of statement of the problem that the number of equations is the required number." He footnotes this remark with the comment, "This apparently is not a unanimous opinion among economists." His footnote goes on to state that Henry Schultz, in reviewing Evans's own book, smuggles integrability into the assumptions of the problem. In terms of comparing states, Evans asks, referring to Schultz's argument, "How many individuals, for instance, can decide, without reference to process, which of the two situations he desires—peace, or justice, in China?" Evans concludes this discussion of utility by saying that if we are to "distinguish between cooperative and competitive elements in the system [we] . . . have already . . . grouped utility indices . . . and these have no transparent relation to the individual ones . . . and from this point-of-view the doctrine of laissez faire lacks mathematical foundation" (324).

The argument winds up with Evans's question, "Would it not be better then to abandon the use of the utility function, and investigate situations more directly in terms of concrete concepts, like profit and money value of production, in order to take advantage of the fact that money is fundamental in most modern economies and to use the numbers which it assigns to objects? Concrete concepts suggest concrete hypothesis" (324).

Thus by 1932, after his book, and after economists have had a chance to respond to the arguments of his book, Evans is unrepentant. Economists, especially mathematical economists of the neoclassical variety, have it wrong. Utility theory, and subjective value theory, founders on the integrability problem. One can only get out of the theory what one puts into it. Reprising Volterra's critique of Pareto more than a quarter century earlier, for Evans, in mathematical economics one should not be so concerned with the behavioral theories themselves. Economic theory, or at least mathematics as applied to economic theory, should trace the implications in logical systems of various hypotheses which themselves are grounded in quantifiable objects or concepts, and the implications are, or should be developed to be, themselves testable either empirically through data analysis or through common sense.

For Evans, as for Volterra, the issue was not formalist versus informalist or antiformalist mathematics, but rigorous versus nonrigorous mathematics. Evans sought rigor in mathematical economics in the way that Volterra had: the mathematical models are not free but are rather tightly constrained by the natural phenomena themselves that those mathematical constructions must model.[10] Evans's image of mathematics looked back, through Volterra, to the optimism of the turn-of-the-century solutions, solutions that were to be abandoned by mathematicians later, to the great challenges faced by mathematics in dealing with set theory, and to that same mathematics in interpreting relativity and quantum phenomena. The move to axiomatics, well under way within the mathematics community by the 1930s, and instantiated in economic argumentation by mathematical economists by the 1940s, left no place in economic theory for Griffith Evans. It does, however, leave an alternative place for Evans. As one of the founding members of the Econometric Society in 1932, Evans subscribed to the call to "promote research in quantitative and mathematical economics . . . [in order] to educate and benefit its members and mankind, and to advance the scientific study and development . . . of economic theory in its relation to mathematics and statistics" (Christ 1952, 5, 11). The point is that Evans's views on mathematical modeling are the views of an econometrician or applied economist today, or one who insists that the assumptions and conclusions of an economic model, a model constructed and developed mathematically, must be measurable or quantifiable. This is the distinction between "modelers" (or "applied economists") and "theorists" that divides modern departments of economics even as both groups consider themselves to be neoclassical economists. That Evans's first important student was Charles F. Roos, one of the early luminaries in econometrics, and founder of his own Econometric Institute in New York (Fox 1987), should allow us to reframe the idea of Evans's "marginalization": it was not that Evans abandoned canonical mathematical economics, but rather that mathematical economics, increasingly connected to the new (very un-Volterra-like) ideas of mathematical rigor in both mathematics and applied mathematical science, moved away from Evans.[11] The image of mathematics, with respect to rigor, had changed. In a real sense, the distinction between rigor as materialist-reductionist quantification and rigor as formal derivation, a distinction contested at the end of the nineteenth century but which disappeared as formalism took hold in mathematics, re-established itself in the distinction between econometrics and mathematical economics, between applied economics and economic theory. It is not unreasonable then to see Lawrence R. Klein as linked to Griffith C. Evans.[12] And I, as Klein's student, unravel the links.

3   Whose Hilbert?

In the Sahara desert there exist lions. Devise methods for capturing them . . . 1. The Hilbert Method. Place a locked cage in the desert. Set up the following axiomatic system: i) The set of lions is non-empty. ii) If there is a lion in the desert, then there is a lion in the cage. Theorem 1: There is a lion in the cage.
—J. Barrington, "15 New Ways to Catch a Lion"

As for the term "formalist," it is so misleading that it should be abandoned altogether as a label for Hilbert's philosophy of mathematics.
—W. Ewald, *From Kant to Hilbert*

Modern controversies over formalism in economics rest on misunderstandings about the history of mathematics, the history of economics, and the history of the relationship between mathematics and economics. More specifically, there is widespread confusion about the nature of and interconnections among "rigor," "axiomatics," and "formalism," as well as "mathematics." For example, when an economist speaks of the connection between mathematics and economics, what is meant by mathematics? Is one concerned with the corpus of mathematical theorems, propositions, and definitions? Or is mathematics rather the totality of views about the methods and construction of such cultural products as those views have developed within the community of mathematicians? Is it the case that

Formal=abstract=axiomatized=mathematical?

If so, then one has taken a particular position with respect to mathematics. To add to that equation "=rigorous economics" is to take views simultaneously on both disciplines.[1]

In this chapter, I will reconstruct a variety of perspectives about formalism, and attempt to untangle the skeins of conflicting meanings. The discussion, as it develops, will necessarily require attention to the history of mathematics, and to some contentious issues in that discipline. We shall see that the confusions expressed by economists, theorized by methodologists, and narrated by historians of economics are collectively intertwined with real disagreements among 1) mathematicians, 2) historians of mathematics, 3) philosophers of mathematics, and 4) historians of the philosophy of mathematics concerning formalism in mathematics.

Fearing Formality

Some economists have been greatly exercised about the connection between economics and formalism. For example, methodologist-economist Terrence Hutchison (1977) in "The Crisis of Abstraction" argued that "contrary to the traditional aims and claims of the subject . . . much of the most highly-regarded work in economics [does not make] . . . and did not seem at all designed to make, any useful contribution to increasingly pressing real-world problems" (68). He goes on to cite economist G. D. N. Worswick, who tells us that "there now exist whole branches of abstract economic theory which have no links with concrete facts and are almost indistinguishable from pure mathematics. . . . Too much of what goes on in economic and econometric theory is of little or no relevance to serious economic science" (70). Hutchison continues to quote, at length, from venerable economists like Ragnar Frisch, Henry Phelps Brown, Wassily Leontief, Nicholas Kaldor, and Harry Johnson, all of whom were distressed by the abstract and mathematically sophisticated modes of argumentation in economic and econometric theory, which, in their collective indictment, impoverish economic discourse.

In another such discussion, in chapter 10 of *Knowledge and Persuasion in Economics*, Deirdre McCloskey takes on the "rhetoric of mathematical formalism" (1994). McCloskey is not entirely hostile to the use of mathematics in economics: "Economics made progress without mathematics, but has made faster progress with it. Mathematics has brought transparency to many hundreds of economic arguments. The ideas of economics—the metaphor of the production function, the story of economic growth, the logic of competition, the facts of labor-force participation—would rapidly become muddled without mathematical expression" (128). Yet McCloskey suggests that if economics really wished to model itself on physics, it would use less, not more, formal mathematics. She believes that reasoning chains in theoretical physics, though constructed with often quite sophisticated mathematics, are themselves moved along by physical argumentation. In her view, in physics unlike economics, mathematical formalism does not constrain argumentation: "The economists, to put it another way, have adopted the intellectual values of the math department—not the values of the departments of physics or electrical engineering or biochemistry they admire from afar. . . . The economists are in love with the wrong mathematics, the pure rather than the applied" (131–32, 145). McCloskey seems to think formal=abstract=pure, making a distinction between pure and applied mathematics, and using this disjunction to say that economists should keep to "good" applied mathematics.[2] Her argument thus concerns the economics-appropriate body of mathematical knowledge (hill-climbing algorithms versus the Hahn-Banach theorem?). At the root of McCloskey's discomfiture is her failure to recognize that what constitutes a good mathematical argument, or a convincing mathematization of a physical theory, has changed over time.

Another discussion appears in Henry Woo's (1986) essay on formalization in economics. In his first footnote, Woo defines his subject:

> "Formalization" is very often used interchangeably with "mathematization" by the author though the former has a broader meaning. Throughout this work, this term is also used interchangeably in some instances with "axiomatization," though again it includes the latter in its meaning. Strictly speaking, formalization encompasses both the syntactical techniques of axiomatization and the semantic techniques of model theory. (20)

Woo discusses, under the rubric of formalization, questions of the nature and role of mathematics, the nature and role of the axiomatic method, and what logicians refer to as formal models. His conclusion gives the flavor of both his puzzlement, and a reader's frustration:

> Formalization is difficult to apply risklessly, and it could be costly if a bad formalization results. Even when these hurdles are overcome,

what formalization can contribute would still be very meager, unimportant, and uninteresting. In other words, formalization in the realm of social sciences is unable to promise much, pays very little, is very costly to conduct, and very risky to contain. Thus, while formalization has revolutionized the development of the physical sciences, we may have come by now to a full circle, where we will find that revolutions in the social sciences have to await the abandonment of the formal method as the chief tool of inquiry. (98)

In a discussion of some of these same issues, Ken Dennis's (1995) critique of formalism in economics similarly conflates formalization with its meaning in metamathematics: "Contemporary mathematical economics is lacking logical rigour because its formal mathematical apparatus neither captures nor expresses the economic content of the theory, and the economic content of the theory is lacking in explicitly formal means of expression by which it can be rigorously set forth and critically inspected" (185).

His argument is that the applied mathematics that is used in economics, which expresses theoretical propositions in economics, lacks logical rigor; for Dennis, an argument is rigorous if and only if it proceeds from assumptions to conclusions at every step satisfying, explicitly, the rules of formal logic. It is well known that no interesting mathematical proof is rigorous in this sense. Dennis goes on to present Adam Smith's invisible hand doctrine from *The Wealth of Nations* and attempts to reduce its informal argument to a set of economic propositions. Then he reframes those propositions as a sequence of formal logical propositions that eventually state that "every x is such that if x intentionally seeks the interest of x, and x does not intentionally seek the general interest, and x competes freely in an open market system, then x unintentionally promotes the general interest" (192). He then goes on to argue that no formalization in these terms does adequate justice to Smith's idea. "The most we find in advanced theory is an informal translation of some of the individual symbols making up the mathematical formulae, without full translation of the formulae themselves. Careful inspection of the mathematical texts of current economic literature will show that most of the non-mathematical content of economic theory is not clearly or explicitly incorporated into any formal apparatus" (198).

What is needed, Dennis concludes, is more self-conscious formalism in the sense of logical formalism, if formalism itself is thought to be desirable in economics:

> Logical form transcends mathematical form, indeed transcends set theory and extensional logic in general. Economic logic springs from language, not from mathematics. To understand economic logic, one must begin with natural language as a datum and make sense of it. . . . Only through the study of economic discourse can we unravel the complexities of rational thought and the rational behavior that follows from it. (198)

From distinguished economists, and economic methodologists, we have a narrative of complaint. The rhetorical strategy involves assailing one's scientific opponents, particularly if they employ new and sophisticated mathematics, for representing empty or overly abstract or arid or unworldly or unrealistic or historically uninteresting modes of inquiry. These attacks themselves, even within economics, constitute a distinct genre.[3] We may, without doing a grave disservice to those individuals who are on record on the subject, call this talk or presidential address or curmudgeonly article "the Mark X version of 'Those Were The Good Old Days,'" or "When Mathematically Unsophisticated Giants Walked the Earth." These discussions are composed equally of sections of mathematical misinformation, piety to a past that never existed, derision of those who would lead the young astray, and professional self-congratulation for having fought the good fight against those barbarians at the gates (or as an alternative, a section of mea culpa: "How I Used to be a Mathematical Barbarian But Then I Saw the Light").

Behind the noise and the posturing, the issues are far from clear. What is this specter called formalism, and why are so many economists concerned that it is somehow detrimental to the production of good economic analysis? My own answer will of course require attention to arguments and reconstructions in the history of mathematics. Since the mathematician David Hilbert is often associated with formalism in mathematics, and several historians of economics have tried to trace formalism in economics to Hilbert's influence on others, it is Hilbert to whom we must first attend.

Should Economists Care About David Hilbert?

The question of David Hilbert's influence on modern economics rests, it appears, on several separate and influential discussions in the history of economics. First, we have the important volume by Ingrao and Israel (1990), which intertwines general equilibrium theory with the history of the mathematization of economics; second, we have Punzo's (1991) argument that reconstituted the Vienna of von Neumann and Morgenstern in terms of what he called Hilbert's Formalist Program; and third, we have some of Mirowski's papers (e.g., 1992), which intertwine formalism, rooted in Hilbert's ideas, and responses to Gödel's Theorem, which ended some of the formalist inquiry, to the emergent work in the theory of games. Related to these papers is Louise Ahrendt Golland's (1996) critique of Punzo's and Mirowski's papers, arguing that those authors were ignorant of and misleading about the relevant history of metamathematics. However, since the Mirowski and Golland concerns can only be addressed after we have examined the mathematical issues, let us here simply introduce Hilbert through the uses (and I shall argue that they were in fact misuses) that historians of economics have made of him.

Lionello Punzo's interpretation of Hilbert's emphasis on axiomatization, and his reading of the record of mathematical formalization, presents a narrative of influence from Hilbert through von Neumann, to the Mengerkreis of Vienna and thus to Abraham Wald, and thence through to proofs of existence of competitive equilibrium models by Debreu and his Hilbert-influenced mathematical teachers among the Bourbaki.

> General equilibrium did not evolve into a metatheory as a result of historical accidents, as Debreu suggests (1984). The evolution was rather the outcome of a sequence of logically connected events which were part of a still semi-articulated or embryonically conceived plan. This plan aimed at redesigning general equilibrium from its very foundations. . . . [In] the program of redesigning economics initiated in Vienna, the use of mathematics as a tool to attain, at least in principle, exact measurability and quantitative predictability of the values of economic variables yielded to the logical calculus. . . . [This revolution, or what mathematically is termed a] catastrophe in the history of eco-

nomic analysis is essentially the result of the influence of the mathematical philosophy called mathematical formalism that was heralded by Hilbert from his stronghold, Göttingen. This philosophy provided two pillars: the axiomatic method, and the principle of hierarchical interdependence between a plurality of theories and the unifying metatheory behind them. (Punzo 1991, 3–5)

In coarse outline, Punzo proposes: 1) Hilbert created a mathematical viewpoint called "mathematical formalism"; 2) This viewpoint permeated Menger's Vienna, particularly the work on general equilibrium models done by Schlesinger, Wald, and von Neumann; 3) general equilibrium theory, in the hands of the Vienna formalists, became a metatheory, and so particular general equilibrium "models as formal abstractions were in need of validation. This was sought in proofs of consistency, because only consistent models would be able to explain their endogenous variables" (16). For Punzo, what he termed "Hilbert's Formalist Program" defined the concerns of the Mengerkreis, and thus formalism came into economics in Vienna in the 1930s through the proofs of the existence of a competitive equilibrium done by Wald and von Neumann. A consequence of this argument is that general equilibrium theory, and thus neoclassical economics itself, has been shaped and formed by Hilbert's mathematical formalism, the midwife of its Viennese birth.

For Ingrao and Israel, although Hilbert's work on geometry "represented the programmatic manifesto of the axiomatic movement . . . whereby a mathematical theory is nothing more than a complex of theorems obtained through deductive logic and defining the properties of a mathematical entity defined by axioms . . . [yet] Hilbert cannot be said to have completely accepted the developments that were the extreme consequences of his viewpoint" (183–84).

They go on to argue that it was Hilbert's young associate John von Neumann who pushed the new ideas whereby the "old deterministic mechanistic viewpoint [was replaced] with one based on the idea of mathematical analogy. The old reductionism was replaced by neoreductionism, whose key idea was the centrality of mathematics, understood as a purely logicodeductive schema" (185). Consequently, it was the von Neumann perspective that shaped general equilibrium theory and game theory, and thus reconstituted economic theory. Thus, David Hilbert was the spiritual grandfather of this new economics.

Mirowski's use of Hilbert is also clear and direct. In a conference paper, Mirowski (1992) developed a narrative of von Neumann's changed approach to the theory of games between von Neumann's early paper (1928) and his 1944 book with Oskar Morgenstern. The Mirowski chronology contextualized that earlier paper by other late 1920s papers done in association with the Hilbert program of axiomatizing various subfields of science. He goes on to argue, though, that following Gödel's 1930 announcement of the incompleteness proof, portions of the formalist program had to be abandoned. Mirowski argues that von Neumann, to replace the certainty of proof, moved to a position that could be described as "strategic," conceptualizing mathematics as a game played by specified rules. For Mirowski then, Hilbert's formalism, based on axiomatization, was ultimately "limited" by Gödel. In that newly limited form it was taken up by von Neumann, whose book on game theory (with Morganstern) reflected this set of mathematical redirections. Mirowski (2001) refines this chronology to argue that after around 1944 von Neumann's interests shifted to computation and thus he turned his back on both formalization and axiomatization.

Although I believe that Punzo's story is somewhat misleading, a careful historical reconstruction of the history of the mathematical community can reshape Punzo's argument in a productive fashion. The Ingrao-Israel analysis is sophisticated, carefully drawn, and attentive to some of the nuances of the history of mathematics itself. Although I shall not quarrel with its main propositions, I shall argue that their self-conscious focus on general equilibrium theory constrains the potentially more significant set of arguments about the interconnection between mathematics and economics. After all, economists entirely ignorant about general equilibrium theory were concerned with, and attentive to, the role of mathematics in economics. Mirowski's argument, intertwining game theory with the changed nature of the formalist program after 1930, requires more attention to the transformation of the Hilbert "program" than to its life.

Since these authors—Punzo, Ingrao and Israel, Mirowski—have been criticized by both historians and economists, and since those criticisms are mutually inconsistent, I will have to spend a bit of time on the life of formalism. As we shall see, the subject is unsettled among historians of mathematics, and as it is connected to the question of what is "modern" about modern mathematics, the stakes are high.

The Hilbert Program, or Not

As set out in earlier chapters, the canonical view of the history of mathematics appears to be that "around 1900" the various crises of foundations (i.e., the paradoxes in set theory, logic, arithmetic) combined to produce a search for a way out of, or around, the difficulties. As an exemplar of this standard line, Golland (1996) tells historians of economics that "a short time after Hilbert's [1900] talk a cloud came over the axiomatization program with the discussion of the paradoxes of set theory evoked by Bertrand Russell's *Principles of Mathematics* (1903) and Gottlob Frege's postscript to his *Grundegesetze der Arithmetik* (1903). . . . Hilbert's formalism was developed in 1918 and 1922 in response to the paradoxes" (2).

Intuitionism, logicism, and formalism were the leading schools which emerged to address problems in the foundations of mathematics, and metamathematicians, logicians, and philosophers of mathematics consider Hilbert to have been the leader of what they called "the Formalist Program." However, I must point out that most of the individuals who have written about mathematical formalism have written from the perspective of logic, or the paradoxes of set theory. The standard chronology that produces formalism and intuitionism as separate frameworks for considering the foundations of mathematics begins with Frege, Cantor, and Peano and then moves to Russell and Whitehead. There is then a jump in the story line (*vide* Golland) to Hilbert's 1918 "Axiomatische Denken," which evolves by 1930 to Gödel's proof. It is difficult for an outsider to gain a critical perspective on how this canonical narrative developed. Nevertheless, it is well known to historians of mathematics that such ideas of formalism versus intuitionism are connected to themes in metamathematics or the philosophy of mathematics. Moreover, those ideas have been historicized with reference to logic and metamathematics and not to the larger discipline of mathematics.

For example, two mathematicians (Vinner and Tall 1982, 753) tell us that "the Formalist Program, known as the Hilbert Program . . . was a call to

mathematicians to prove the consistency of mathematics within the restricted framework of mathematics," while another (Dawson 1979, 740)
informs us that "Gödel's Incompleteness Theorem . . . shattered the hopes
of those committed to Hilbert's formalist program." Meanwhile, philosophers (Barwise and Moravcsik 1982, 212) tell us that "to logicians, the word
'formal' is strongly associated with Hilbert's formalist program" and metamathematicians (Kleene 1976, 767) affirm that "Gödel's paper showed that
Hilbert's formalist program could not be carried out in any simple way."

Hilbert's "Axiomatische Denken" has not usually been contextualized
outside of the logical-mathematical history (but see Corry 1997, below).
However, it is not clear how that standard account can make sense of the
larger corpus of mathematical work. That is, this usual perspective seems to
account reasonably well for the battle as seen by the metamathematicians,
and philosophers of mathematics, but is inconsistent with the emergent
developments in the rest of mathematics. As Stump puts it, "The development of logic and of the foundations of mathematics has been seen very
much through [Bertrand] Russell's eyes (or perhaps through the eyes of
later standard interpretations of Russell), and for Russell it does often appear that epistemological issues are central" (388). Do we have here a curious doppelganger to our earlier discussion in the first chapter of the self-
referential Cambridge tradition in economics? Is this another example of
how the Cambridge view, represented by Bertrand Russell, a narration centered on Cambridge interests, has stabilized claims about intellectual history, in this case of mathematics? The view that histories are written by
those who have a stake in the outcomes means that historians of mathematical practice have ceded the Hilbert turf to historians of the philosophy
of mathematics. These scholars professionally contest the nature of mathematical objects, theories, and proofs, and their canonical histories portray
Hilbert and Göttingen as epistemologically and ontologically obsessed.

A Different Hilbert

Fortunately for this narration of the history of mathematics in economics,
in recent years some historians of mathematics have reconstructed the
metamathematicians' Hilbert. For example, consider the Hilbert of Ewald

(1996), who argues that "Hilbert is persistently misconstrued as a 'formalist,' i.e., as someone who was so shaken by the paradoxes that he took up the theory that mathematics is merely a game played with meaningless symbols. But the intellectual background of Hilbert's proof theory was richer than this. . . . Hilbert viewed formal axiom systems instrumentally, as a powerful tool for mathematical research, a tool to be employed when a field had reached a point of sufficient ripeness" (1106–7).

To further complicate the usual story such as that presented by Golland, we have a rejection of her chronology by the historian of mathematics Leo Corry (1997). He places Hilbert's concerns *with axiomatization* clearly and firmly with the axiomatization of physics, and not with the paradoxes of set theory. This is not to say that Hilbert was unconnected to the emergent foundational issues of arithmetic. After all, his *Grundlagen* (1899), in particular in the second edition of 1903, was in part concerned with possible contradictions in Euclidean geometry. As Corry notes, for Hilbert the question of the consistency of geometry was reduced to that of the consistency of arithmetic since any contradiction existing in Euclidean geometry must manifest itself in the arithmetic system of real numbers. Thus, among the 1900 list of twenty-three problems the second one concerns the proof of the compatibility of arithmetical axioms (119–23).

Nevertheless, it is not just Hilbert's second problem that is of issue. His sixth problem, the axiomatization of physics, was presented simultaneously. Thus a chronology that develops the historical context as set theory, leaving out physics, produces some potentially serious misreadings.[4] For Hilbert, let us remind ourselves, was a professor at the Mathematical Institute at Göttingen, and his lectures on mathematical physics were closely connected to all of the emerging problems in physics at the turn of the twentieth century. To give the flavor of the physics background of the kinds of problems that Hilbert was to discuss under the heading of "axiomatization," Corry introduces us to Paul Volkmann (1856–1938): "In the intimate academic atmosphere of Konigsburg [where Hilbert began his career], Hilbert certainly met Volkmann on a regular basis, perhaps at the weekly mathematical seminar directed by Lindemann" (101).

Corry notes that Volkmann himself went through a number of different changes in his views on the foundations or first principles of physical theories. However, by 1900, Volkmann would write that the conceptual system of physics should not be conceived as one which is produced bottom-up like a building. Rather it is a thorough system of cross-references, which is built like a vault or the arch of a bridge, and which demands that the most diverse references must be made in advance from the outset, and reciprocally, that as later constructions are performed the most diverse retrospections to earlier dispositions and determinations must hold. Physics, briefly said, is a conceptual system which is consolidated retroactively. (Volkmann 1900, 3–4, as quoted and translated by Corry 1997, 102)

This language of Volkmann will reappear in Hilbert's 1918 paper, to be discussed in this chapter.

Corry, in tracing the roots of Hilbert's concern with foundational issues, found many clues in the surviving notes of the lectures Hilbert gave. "Sometime in 1894 Hilbert became acquainted with Hertz's idea on the role of first principles in physical theories. This seems to have provided a final, significant catalyst toward the whole-hearted adoption of the axiomatic perspective for geometry, while simultaneously establishing, in Hilbert's view, a direct connection between the latter and the axiomatization of physics in general" (1997, 105). Hilbert's lectures for 1894 contained his remarks that "geometry is a science whose essentials are developed to such a degree, that all its facts can already be logically deduced from earlier ones. Much different is the case with the theory of electricity or with optics, in which still many new facts are being discovered. Nevertheless, with regards to its origins, geometry is a natural science" (translated by Corry 1997, 106–7, from Toepell 1986, 58).

Corry goes on to remark that

it is the very process of axiomatization that transforms the natural science of geometry, with its factual, empirical content, into a pure mathematical science. There is no apparent reason why a similar process might not be applied to any other natural science. In the manuscript of his [Hilbert's] lectures, we read that all other sciences—above all mechanics, but subsequently also optics, the theory of electricity, etc.— should be treated according to models set forth in geometry. (107–8)

It is here that we must call attention to a feature of our story implicit in the several interpretations of Hilbert's ideas already mentioned: historians of mathematics are reinterpreting Hilbert, and formalism, as part of a larger concern with the nature of the break between modern mathematics and what came before. Some sense of the immensity of the historical project may be gleaned from the survey-review article by David E. Rowe titled "Perspective on Hilbert" (1997), which considers three recent books (Mehrtens 1990; Toepell 1986; Peckhaus 1990) all dealing with Hilbert's role in creating modern mathematics. Rowe argues that these books direct attention away from the simplistic reading of Hilbert's legacy as one of "formalism" or "axiomatics." But all have differing readings of Hilbert "as a spokesman for modern mathematics during the [1900–1920] period when his influence was most directly felt" (Rowe 1997, 564). For example, Mehrtens book in particular "marks the first serious attempt to understand the complex process of modernization that rapidly transformed both mathematical productivity and the nature of mathematical knowledge during the early decades of this century . . . [while it also] spotlights the role of the historian as a mediating agent between present and past, an activity that takes place in its own rich context of symbols and meanings" (546, 565).

I have argued, in chapter 1, that English mathematics around 1900 was still a mixture of geometry and applied mechanics, with inconsistent images of mathematical truth located in both logic and nature, and have alluded to the fact that this was certainly not so in the other European countries, or what was then referred to as the "Continent." The France of Cauchy, Dirichlet, and Poincaré, the Italy of Peano and Volterra, the Sweden of Mittag-Leffler and Lie, and most of all the Germany of Klein, Weierstrass, Cantor, and Hilbert had been establishing new expectations about proof, and new ideas of rigor based on a more self-conscious connection between the foundations of mathematics, set theory, arithmetic, and logic. It was therefore not a revolution that was heralded by David Hilbert in his lecture, "Axiomatisches Denken", delivered at a meeting of the Schweitzerische Mathematische Gesellschaft in Zurich in 1917, but a point of view that was fully understood, and well represented in mathematical work, by serious mathematicians.[5]

Nevertheless, by 1917, even the English mathematics community understood that the older view of mathematical truth had collapsed. In the same year of Hilbert's talk, Alfred North Whitehead noted, "The whole apparatus of special indefinable mathematical concepts, and special a priori mathematical premises, respecting number, quantity and space, has vanished" (Whitehead 1917, 361). In other words, by the time Hilbert was writing on the new axiomatic way of thinking, writing about mathematics and axiomatics, even the previously backward English mathematical community was beginning to change.

The reasons for that change remain problematic. David Stump (1997) has recently argued that

> much of the work on foundations and the formalization of mathematics—the axiomatization of mathematical theories, the use of explicit definition, and the reduction of the number of primitive terms and of axioms to a minimum—was in response to the problem of growth and specialization in mathematics during the nineteenth century. Many mathematicians searched for unifying basic concepts and attempted to reduce the number of primitive terms and axioms to a minimum. . . . [Against this view the] standard accounts of the history of the foundations of mathematics claim that the investigation of foundations was motivated by the discovery of paradoxes and set theory, a story of crisis that leads one to believe that the primary motivation for the study of the foundation's of mathematics was epistemological and originated in response to doubts about the consistency of mathematics and the truth of some of its branches. (383–84)

For historians of economics attempting to sort out the interconnections between changing views of mathematics and a changing mathematical economics, the lack of agreement on the nature of the changes in mathematics—witness Richards, Stump, Corry, Mehrtens, Rowe, et al.—can be unsettling. Nevertheless, a direct archaeology of the historiography would take us too far afield.

"Axiomatisches Denken"

If we simply lay aside these issues, recognizing that historians of mathematics, as opposed to historians of the philosophy of mathematics, write different histories of formalism, and construct differently interested David Hilberts, we can look directly at Hilbert's 1917 talk (1918) concerning the role played by axiomatization. He set out the main point quickly: in a variety of mathematized fields like elementary radiation theory, like mechanics, like thermodynamics, like the theory of equations, there are a small number of crucial propositions that characterize the field.

> These fundamental propositions . . . can be regarded as the axioms of the individual fields of knowledge: the progressive development of the individual field of knowledge then lies solely in the further logical construction . . . of the framework of concepts. This standpoint is especially predominant in pure mathematics . . . [although] in the [applied] fields of knowledge the need arose to ground the fundamental axiomatic propositions themselves. (Hilbert [1918] as translated in Ewald 1996, 1108–9)

The central building blocks, or propositions, or theorems, of many theories—like the laws of arithmetic and the principle of entropy—themselves are often capable of being further grounded in anterior propositions, and this process may be continued. This is the procedure that led, for example, to the development of Peano's postulates for arithmetic, and Russell and Whitehead's axiomatization of logic. In the central metaphor of the argument, Hilbert uses the idea of constructing a building (recall Volkmann) by developing foundations, scaffolding, and so on.

> The procedure of the axiomatic method, as it is expressed here, amounts to a deepening of the foundations of the individual domains of knowledge—a deepening that is necessary for every edifice that one wishes to expand and to build higher while preserving its stability.
>
> If the theory of a field of knowledge—that is, the framework of concepts that it represents—is to serve its purpose of orienting and ordering, it must satisfy two requirements above all: first it should give us an overview of the independence and dependence of the propositions of the theory; second, it should give us a guarantee of the consistency of all the propositions of the theory. In particular, the axioms of each theory are to be examined from these two points of view. (1109)

The call for axiomatic thinking, for formalizing wherever possible a theory or linked set of theories in a moderately well-defined scientific discourse, is clearly laid out in Hilbert's talk. One tries to identify a major proposition, or theorem, or building block of the subject. One then tries to develop that theorem from more fundamental propositions by asking how the theorem might be derived, or proved, if one had the more fundamental propositions or axioms at hand. One then continues in this fashion, sinking concrete pillars into the muck, as it were. At some point one stops, perhaps because the propositions are no longer reasonable, or physically tractable to interpretation. One then asks if the axioms one is left with are independent and consistent. Independence means that each axiom is neither derivable from, nor can be used to establish or prove, any other axiom. Consistency of a set of axioms means that there is no contradiction to be produced in the theory by assuming the truth of the set of axioms or fundamental propositions. This may be assessed by asking if there is at least some model, or physical interpretation of the set of axioms, such that all members of the set are true under that interpretation or model, for if that is the case, then there is no logical contradiction that can arise, no theory based on those axioms will contain an internal contradiction. As an example, Hilbert notes that "modern quantum theory and our developing knowledge of the internal structure of the atom have led to the laws which virtually contradict the earlier electrodynamics, which was essentially built on the Maxwell equations; modern electrodynamics therefore needs—as everybody acknowledges—a new foundation and essential reformulation" (1112).

What we have then is less a concern that mathematics itself be free of paradoxes in the first instance, as much as a set of suggestions for developing and constructing theories. Hilbert's 1918 argument, if taken as a manifesto, is thus as much a call to change what constitutes mathematical knowledge, and to modify the ways mathematics views its own claims, as it is a call for specific mathematics to be created, and thus for the body of mathematical knowledge to be augmented in specific ways.[6] It is a program not only for establishing knowledge, but also for organizing and supervising the growth of knowledge itself. The process of axiomatic thought is then a method both for accreting and warranting knowledge claims, for those claims, if developed from independent and consistent axioms, themselves make strong claims on our attention and reason. Hilbert's call was to reconstitute the image of mathematics.

Let us be clear that Hilbert made the strong claim that formalization *in this sense of axiomatization* was to be pivotal in new scientific work.

> If we wish to restore the reputation of mathematics as the exemplar of the most rigorous science it is not enough merely to avoid the existing contradictions. . . . [Our theory] must go farther, namely to show that within every field of knowledge contradictions based on the underlying axiom-system are absolutely impossible. . . . [Thus] I proved the consistency of the axioms laid down in the *Grundlagen der Geometry* by showing that any contradiction in the consequences of the geometrical axioms must necessarily appear in the arithmetic of the system of real numbers as well. (Ewald 1996, 1112)

This project thus requires one to establish the consistency of systems in terms of the consistency of the integers, or arithmetic, and that of set theory itself. This was the Frege program, carried out by Russell and Whitehead for logic. They showed that logic of a certain type was in fact consistent in terms of a certain kind of set theory. Hilbert noted that their result forced the special questions of the consistency of set theory, and arithmetic, out into the open. Further, it meant that the open questions of the consistency of set theory and arithmetic were fundamentally linked to 1) the problem of the solvability in principle of every mathematical question; and 2) the problem of the nature of a mathematical proof, specifically the idea of the decidability of a mathematical question by a finite number of operations.

It is in this sense that Hilbert laid out a research program, the program of axiomatizing mathematical, or applied mathematical, or physical, or even social, theories. This program is not a call for rigor as opposed to intuition in mathematics, nor did it call for a change in the way mathematics was henceforth to be done, but rather it sought a method for organizing and systematizing mathematical theories.

In a fine appreciation of Hilbert, the mathematician Herman Weyl wrote that

> Hilbert is the champion of axiomatics. The axiomatic attitude seemed to him one of universal significance, not only for mathematics, but for all sciences. His investigations in the field of physics are conceived in the axiomatic spirit. In his lectures he liked to illustrate the method by examples taken from biology, economics, and so on. The modern epistemological interpretation of science had been profoundly influenced by him. Sometimes when he praised the axiomatic method he seemed

to imply that it was designed to obliterate completely the constructive or genetic method. I am certain that, at least in later life, this was not his true opinion. For whereas he deals with the primary mathematical objects by means of the axioms of his symbolic system, the formulas are constructed in the most explicit and finite manner. In recent times the axiomatic method has spread from the roots to all branches of the mathematical tree. Algebra, for one, is permeated from top to bottom by the axiomatic spirit. One may describe the role of the axioms here as the subservient one of fixing the range of variables entering into the explicit constructions. But it would not be too difficult to retouch the picture so as to make the axioms appear as the masters. An impartial attitude will do justice to both sides; not a little of the attractiveness of modern mathematical research is due to a happy blending of axiomatic and genetic procedures. (Weyl 1944, 274)

A New Reductionism

Thus by the end of the second decade of the twentieth century Hilbert and other mathematicians were developing mathematical structures useful to provide models for applied fields (Hilbert mentions electrodynamics, radiation, thermodynamics, gravitation, quantum theory, etc.). Replacing the late-nineteenth century reductionism of modeling phenomena by mechanical structures, such research was creating a framework for a mathematical reductionism. That framework reflected fundamental concerns in the foundations of mathematics itself. This program thus had a number of elements to it.[7] At its most applied level, it required axiomatization of particular scientific theories. More fundamentally, it asked that the axiomatization be consistent in terms of systems that are more fundamental. Consequently, the program required special attention to the axiomatization of both set theory, and arithmetic, for the consistency of those systems was the basis of attention to the consistency of systems based on them. The real numbers, for example, could be built up by taking the integers and adding the Dedekind axiom. If arithmetic was consistent, then so too would be the real numbers. Now if one has a consistent system, a particular proposition expressible in that system will be either true or false (in that system) else that system is not complete. Of course, one could always add as an axiom to the system that proposition which is neither true nor false in the system and thus make the system more complete. The completeness of a system thus is tied to the problem of the decidability of propositions, or the possibility of mathematical proof itself.

Consequently, both the consistency and completeness of set theory, and arithmetic, were fully on the research agenda for mathematics, as outlined by Hilbert in his various writings and talks, as the historians of the philosophy of mathematics have argued. If it could be established that arithmetic and/or set theory, or more precisely the particular axiomatizations of arithmetic and/or set theory, were both consistent and complete, then mathematization settled the epistemological quest for certainty. This gives us a strong reading of Hilbert's concluding paragraph of "Axiomatisches Denken":

> I believe: anything at all that can be the object of scientific thought becomes dependent on the axiomatic method, and thereby indirectly on mathematics, as soon as it is ripe for the formation of a theory. By pushing ahead to ever-deeper layers of axioms . . . we also win ever-deeper insights into the essence of scientific thought itself, and we become more conscious of the unity of our knowledge. In the sign of the axiomatic method, mathematics is summoned to a leading role in science. (Ewald 1996, 1115)

This was the call, which ends dramatically by reminding the hearer-reader of Emperor Constantine's "in hoc signe vinces." This was the vision of the clarity and rigor that formalization and axiomatization would bring to scientific practice.

Hilbert's charge to mathematics thus had two parts: one called for a change in mathematical knowledge (attention to issues of proof theory, for example), and the other called for a change in the image of mathematics (whose task became that of providing a rich store of postulational or axiom systems, and mathematical structures, to sustain the mathematical research community). Following a suggestion made to me by Leo Corry, I will call the first Hilbert's "Finitistic Program for the Foundations of Arithmetic" (FPFA) and the second Hilbert's "Axiomatic Approach" (AA) (not "Formalist Program").[8] It is my contention that only the latter was to play a role in the development of a mathematical economics.

It is a historiographic problem that the FPFA has been taken to define something called "Hilbert's Formalist Program" by historians of the philosophy of mathematics, who have too often been taken as authoritative by historians of economics. In contrast, I insist that it is historians of mathematical practice, like Leo Corry and Giorgio Israel, who should hold our attention for they have, implicit in their perspective, a fascinating story to tell economists. It is to that story I will soon turn. However, one more bit of archaeology on formalism is required for my narrative to cohere, for Hilbert's FPFA met a reversal at Gödel's hands, and consequently ideas about formalism in mathematics underwent some changes that appear to affect arguments about formalism in economics.

Kurt Gödel

Recapitulating the argument to this point, we have seen that a number of economists and historians of economics (e.g., Hutchison, Phelps-Brown, Leontief, most Post Keynesians, Austrians) have argued that traditional economics was essentially subverted in the twentieth century. The vice of the new economics was its increased concern about mathematization and/ or axiomatization. In the view of those economists, this turn was "formalism," and so it could be linked (although not by them) to formalism in mathematics, and the individual identified by historians of metamathematics and philosophers of mathematics as the progenitor of the "Formalist Program," David Hilbert. The argument, in the hands of Punzo, for instance, becomes one of influence whereby Hilbert's ideas influenced von Neumann, whose ideas connected to the Mengerkreis of Morgenstern, Menger, and Wald, and thence resurfaced in Cowles in the United States and in *The Theory of Games and Economic Behavior,* going on to become both the methodology of the Cowles-led Econometric Society as well as the intellectual wellspring of the Arrow-Debreu general equilibrium theory, which made neoclassical economics safe and respectable.

Confusing this position is that it is often joined (e.g., Dennis) to the argument that the Hilbert formalist program was a failure outside economics, because of Gödel's work, and consequently economists who are formalists surely are misguided in their belief that axiomatization is a worthwhile activity for the sciences. Moreover, the excellent historians of the mathematical economics of the twentieth century (e.g., Mirowski, Ingrao and Israel, Punzo) themselves are ambivalent about the intertwining of mathematics and economics. Mirowski argues for instance that mathematical economics (until approximately 1944) was inextricably linked with physical reductionism, so that the neoclassical theory of individual behavior is mid-nineteenth-century energetics in disguise and is incapable of bearing explanatory weight. Finally, attempts to straighten these matters out historically have failed to understand (e.g., Golland) that the history of mathematics is not in fact coextensive with the history of metamathematics. Consequently, we must ourselves look in more detail at the fate of Hilbert's FPFA and AA in Gödel's hands, since I shall argue that the latter survived quite well even after what the historian of mathematics Morris Kline called "the loss of certainty" in mathematics.

Kurt Gödel was born on April 28, 1906 in what was Brunn, Moravia, which later became Brno, Czechoslovakia.[9] A member of the German community there, he attended German-language schools before enrolling at the University of Vienna in 1924. He moved into mathematics in 1926, and began work on the foundations of mathematics under the supervision of the mathematician Hans Hahn. He also attended meetings of the Vienna Circle of the philosophers Schlick, Neurath, Carnap, and Waismann, and mathematicians Hahn and Karl Menger, as well as Menger's own mathematical circle, known as the Mengerkreis.[10] Gödel's dissertation was submitted in the fall of 1929. In it, he established the completeness of the first-order predicate calculus. That is, based on a formalization of first-order logic developed by Hilbert and Ackerman in 1928, Gödel showed that "for each formula A of the first-order predicate calculus, either A is provable or its negation . . . is satisfiable in the domain {0, 1, 2, . . . } of the natural numbers" (Kleene 1988, 49). This result "stands at the focus of a complex of fundamental theorems, which different scholars have approached from various directions" (ibid.). The result was, under all interpretations, a striking positive result for the Hilbert Program.

Gödel went to Königsburg on 7 September 1930 to present the result to a session at a major conference on epistemology, organized by the Gesellschaft für empirische Philosophie. He apparently had told Carnap, when they met in a coffeehouse two weeks earlier, of another result he had just established, and that he intended to mention at the conference. There is some evidence to suggest that Hahn was aware of the other result too.

> At the time, Gödel was virtually unknown outside Vienna; he had come to the conference to deliver a twenty minute talk on his dissertation, completed the year before and just then about to appear in print. In that work, Gödel had established a result of prime importance for the advancement of Hilbert's programme . . . so it could hardly have been expected that the day after his talk Gödel would suddenly undermine that programme by asserting the existence of formally undecidable propositions. (Dawson 1988, 76)

Specifically, Gödel announced that there was a proposition neither true nor false in the formal systems of Russell and Whitehead's *Principia Mathematica* and in the Zermelo-Fraenkel axiomatization of set theory. Gödel's results were set out in the paper "Über formal unentscheidbare Sätze der *Principia Mathematica* und verwandter Systeme I," translated as "On Formally Undecidable Propositions of *Principia Mathematica* and Related Systems I" (Shanker 1988).[11] Consequently, set theory could not be complete. This result produces, in logicians and mathematicians, almost a feeling of awe at its magnificent accomplishment.

> It is natural to invoke geological metaphors to describe the impact and lasting significance of Gödel's incompleteness theorems. Indeed, how better to convey the impact of those results—whose effect on Hilbert's Programme was so devastating and whose philosophical reverberations have yet to subside—than to speak of tremors and shock waves? The image of shaken foundations is irresistible. (Dawson 1988, 74)

As Dawson notes in his survey of what is known about the reception of Gödel's result in the philosophical and mathematical communities, the folklore is that Gödel "presented his results with such clarity and rigour as to render them incontestable" (75). In fact that was not so, and it was a few years before the full impact of the results was understood. Those few years were a time in which Gödel extended the scope of his original paper, giving informal proofs of the propositions like the unprovability of consistency for axiomatic set theory, and the formal undefinability of the notion of truth. He corresponded with, and gave talks to, all the major theorists concerned with the issues: Bernays, Hilbert, Carnap, von Neumann, and Zermelo. "During 1931, Gödel spoke on his incompleteness results on at least three occasions: at a meeting of the Schlick circle [the Vienna Circle] (15 January), in Karl Menger's mathematical colloquium (22 January), and, most importantly, at the annual meeting of the Deutsche Mathematiker-Vereingung in Bad Elster (15 September)" (76).

In any event, without going into too much detail about the manner by which Gödel's results were broadcast to, and assimilated by, the relevant communities, we can be confident that by the middle of the 1930s Gödel's results were beyond question. He was understood to have changed forever the optimistic hope of the Hilbert FPFA that all scientific knowledge could eventually be formalized, and developed axiomatically, on secure foundations in mathematics. Hilbert's FPFA, which called for the production of particular mathematical results which would show the consistency of mathematics, was thus shown impossible. It did not of course dash the optimism of the Hilbert AA, his call to explore foundations and develop theories from axiomatic foundations. Nor did it touch Hilbert's arguments for the introduction of formal rigor wherever possible in scientific work. Gödel's impossibility theorem thus did not touch the change in the mathematician's image of the activity of making mathematical knowledge more secure, and how to pursue new scientific knowledge in an organized and rigorous manner: it did not mute Hilbert's call for "Axiomatisches Denken."

John von Neumann

Let me now break off this line of thought to return to John von Neumann's role in the story. Since I have, in an earlier book (Weintraub 1985), brought together some of the issues of von Neumann's background, I can at present be a bit sketchy on the biographical details, leaving them to be integrated with my other discussion by interested readers.[12] Von Neumann was the potential intellectual heir to David Hilbert. Moreover, von Neumann was fully involved in the mathematical work associated with Hilbert. This is clear from two particular lines. First, we have the evidence of von Neumann's research on the foundations of logic and set theory in the 1920s, directly linked to the Hilbert FPFA. Second, we have the important example of von Neumann's axiomatization of quantum mechanics of

1928. This paper made manifest Hilbert's own view of how scientific theories should be formalizable in the sense that their fundamental theorems should be developed from a formal axiomatic base, what I have called Hilbert's AA.

It was around 1927–28 that von Neumann took up the question of two-person zero-sum games, and provided, in the paper "Zur Theorie der Gesellschaftsspiele," a rigorous proof of the minimax theorem. This paper has some links to thermodynamics since the objective function to be "minimaxed" resembles a potential function. Whether von Neumann came to the game theory propositions by this means, or through some earlier incomplete work by Borel as some French mathematicians have argued, or whether von Neumann's own interest in games was quite independent, is not relevant here.[13] Rather, we note that von Neumann's 1928 treatment of games is developed axiomatically, and thus von Neumann's concern is here again consistent with the approach articulated by Hilbert.

In addition, around this time, von Neumann began work on the paper that was to be published in 1936 as "Über ein Ökönomishes Gleichungssystem und eine Verallgemeinerung des Brouwerschen Fixpunksätzes." I have argued before (Weintraub 1985) that this was the most important paper done in mathematical economics: it was the genesis of 1) modern existence proofs in general equilibrium models; 2) linear programming and dual systems of inequalities; 3) turnpike theory; and 4) fixed-point theory. The "dating" of the ideas brought forward in this paper creates some confusion in people's minds. Since its relation to the collapse of the Hilbert FPFA is one of my concerns here, let me be very specific. This paper, though it certainly began life in the late 1920s, at a time when von Neumann was "practicing" formalization of theoretical systems, like set theory and quantum mechanics, was publicly read apparently for the first time only in 1932 at a Princeton mathematics seminar. It was not to be published until the 1936 volume of Menger's *Ergebnisse Eines Mathematischen Kolloquiums,* which did not appear until 1937. It was not translated into English until it appeared in the 1946–1947 volume of *The Review of Economic Studies.* Thus unless von Neumann was late in accepting the effect of Gödel's results on the Hilbert FPFA, it cannot be argued that the general equilibrium paper belongs to the same group of papers as those late-1920s papers on axiomatizing quantum mechanics and the theory of games. That they are all connected in fact is indisputable. Punzo shows that there are important links among the mathematical and formal structures of the papers (Punzo 1991, 9–11). But they do not reflect the same point of view about the value and potential importance of axiomatization and formalization of scientific theories, for the earlier papers were fully formed before Gödel's results were known, while von Neumann worked out the general equilibrium theory paper only after 1930, and his acceptance of the import of Gödel's results.

Indeed, von Neumann was present when Gödel made his public remarks in that discussion in Königsburg on 7 September 1930: "After the session he drew Gödel aside and pressed him for further details. Soon thereafter he returned to Berlin, and on 20 November he wrote Gödel to announce his discovery of a remarkable [*bemerkenswert*] corollary to Gödel's results; the unprovability of consistency. In the meantime, however, Gödel had himself discovered [this same result which was] his second theorem and had incorporated it into the text of his paper" (Dawson 1988, 77–78).

By late 1930, von Neumann himself was fully aware that the Hilbert FPFA, in the sense of Hilbert's Second Problem, was dead.[14] Additionally, the broader context that supported the development of his own general equilibrium paper, the Menger Colloquium in the Department of Mathematics at the University of Vienna, was run by Menger, who "in the spring of 1932 . . . became the first to expound the incompleteness theorem to a popular audience, in his lecture Die neue Logik" (Dawson 1988, 81). It will also be recalled that participating in the Menger seminars was Abraham Wald, who published the first two of the proofs on existence of a competitive equilibrium in the Menger Ergebnisse series in which von Neumann's paper was eventually to appear. Indeed, in the published discussion of the second, and most important, Wald paper, there appears the following: "Gödel: In reality the demand of each individual depends also on his income, and this in turn depends on the prices of the factors of production. One might formulate an equation system which takes this into account and investigate the existence of a solution" (Baumol 1968, 293).

Thus von Neumann's 1936 paper, the Wald papers, and the activities of the Mengerkreis instantiated Hilbert's Axiomatic Approach. Mirowski is thus quite right (contra Golland) to point out that whatever von Neumann's interests in the 1930s, they have to be contextualized with respect to a world in which Gödel's results were accepted, a world in which Hilbert's FPFA had failed.

Reflecting on these matters, Punzo (1991) claimed:

> In the recent historical evolution of economics a peculiar blend of
> ideas belonging to mathematical formalism has prevailed, a blend
> which looks in many ways like a compromise to cope with Gödel's
> criticism of the formalistic program. . . . In the program of redesigning
> economics initiated in Vienna, the use of mathematics as a tool to
> attain, at least in principle, exact measurability and quantitative pre-
> dictability of the values of economic variables yielded to the logical
> calculus. A model was reduced to a manipulation of essentially sym-
> bolic strings. . . . From Hilbert's own special version of the axiomatic
> approach, modern general equilibrium derives the notion of econom-
> ics as the analysis of formal systems rather than synthetic representa-
> tions of actual economies. (Punzo 1991, 4)

From Hilbert to von Neumann, to the Mengerkreis and Wald, to Bourbaki
and thence to Debreu runs the chain of causality, the development of
modern economic theory in its unconcern to study real economies.

My own argument suggests that Punzo blurred the distinction between
Hilbert's FPFA and AA, and thus has difficulty distinguishing the separate
issues of creating mathematical knowledge and justifying mathematical
knowledge. He consequently plays down the earlier history of general
equilibrium theory itself (Cournot, Walras, Pareto, etc.) that arose from
attempts to mathematize many fields. As we have seen in chapter 1, such
mathematization had its roots in the rational mechanics of the mid-
nineteenth century, in the enthusiasm for energetics and field-theoretic
physics, not in the mathematics of the Erlanger Program of Felix Klein and
thence Hilbert. Put another way, the sequence whose first term is David
Hilbert, and whose nth is Gerard Debreu (but what of Arrow?), does not
converge. It thus needs to be studied in terms of its subsequences, which is
what my own history of these matters seeks to do.

Reconfiguring Economic Knowledge

I have been here arguing that there is little sense to be made of "the Hilbert
Formalist Program." Instead, Hilbert's views involve two intertwined lines
in his own *evolving* research program: first, it can be understood as a quest

for certainty in mathematics, based on the Second Problem of 1900, which sought a proof of the consistency of arithmetic or logic or set theory. I called this Hilbert's Finitistic Program for the Foundation of Arithmetic (FPFA). But second, it can be understood as the quest to develop axiomatic formulations of not only mathematical theories but also scientific theories more generally, along the lines of Hilbert's Sixth Problem of 1900. I called this Hilbert's Axiomatic Approach (AA).

Gödel's paper of 1930 showed that the FPFA could not succeed: it was impossible to be certain that the foundations of knowledge (if that knowledge were to be based on logic or mathematics) were consistent and thus would not lead to contradictions. Nevertheless, the theorem opened up an alternative approach to certainty, namely "relative certainty," since one was often able to show consistency relative to an extended set of postulates or axioms: if a proposition P was undecidable in system A, appending P to A (extending the axiom system) could assure P's truth, as it were: for any system, truth as consistency was to be relative to the structure in which that system was embedded. So, for example, if two-person game theory were to be formalized, it would be as true (i.e., its conclusions would be true if its assumptions were true) as the logic itself could guarantee, as true then as arithmetic. Truth and consistency were thus intertwined, and consistency was established by relating the theory to a "model" known to be consistent, like a physical model. As we saw in earlier chapters, the image of mathematics shared by Volterra, Evans, Edgeworth, and Pareto used mechanical reductionism to make scientific arguments rigorous. Such rigor guaranteed truth by embedding or linking the economic model to a physical model. In contrast, the emerging view of mathematical truth, Hilbert's AA, appeared to require a different conceptualization. This new image of mathematics shaped an emergent mathematical economics. To preserve the relationship between rigor and truth, economists began to associate rigor with axiomatic development of economic theories, since axiomatization was seen as the path to discovery of new scientific truths. Hilbert's research program in this form had indeed found its way into mathematical economics.

As Corry puts the matter,

> The most fitting postulational [axiomatic] systems were sought which could be used for a starting point for coherent research of a particular

mathematical discipline. In such cases the discipline in question . . . was at the center of interest while the system of postulates was only a subordinate tool, meant to improve research in the former . . . But on the other hand, a new autonomous mathematical activity was developing, which focused on postulate systems themselves as an object of inquiry, . . . [which] had a great influence on the development of mathematical logic in America. . . . [Such] postulational analysis was a direct offshoot of the kind of research initiated by Hilbert . . . but an offshoot that followed a direction not originally intended by him and which went much further than Hilbert would have thought mathematically worthwhile. (1996, 179–80)

Pursuit of the issues brought to the fore by the FPFA gave rise to the field of metamathematics. Corry's argument shows that the emergent concern with formalism by historians of philosophy and philosophers of mathematics is rooted in issues of "postulational analysis" quite apart from any connection between mathematics and scientific knowledge. I submit then that historians of economics cannot look to those communities of philosophers to help us understand the developing connection between mathematics and economics in the twentieth century.

Mixing the connection between mathematics and economics with the idea of formalism in economics is explosive for those who try to reconstruct the history of economics in the twentieth century. It is easier to reject "Formalism" than it is to come to terms with the Axiomatic Approach. I suspect that this is the reason why the related topic of the proper role of mathematics in the social sciences is so very controversial. What is at stake is, to put it starkly, the very concept of scientific truth—economic truth— itself.

The concept of a true scientific theory has changed over the twentieth century as the image of mathematical knowledge changed. The FPFA has failed. However, the relativization of scientific knowledge (communally stable beliefs) in the sense of the AA does not mean that there is no scientific "truth" to be obtained in any field. Rather it allows the relevant scientific community to accept claims to knowledge as true, and to embed that knowledge in the practices, language, models, and theories of the community. For mathematicians, acceptance is based on the communally agreed upon idea of a "good" mathematical proof:

> Abstraction, formalization, axiomatization, deduction—here are the ingredients of proof. . . . Proof serves many purposes simultaneously. In being exposed to the scrutiny and judgement of a new audience, the proof is subject to a constant process of criticism and revalidation. Errors, ambiguities, and misunderstandings are cleared up by constant exposure. Proof is respectability. Proof is the seal of authority. . . . Finally, proof is ritual, and a celebration of the power of pure reason. (Davis 1981, 150–51)

If mathematical knowledge is communal and contextual, and mathematical knowledge undergirds scientific knowledge, then the idea of scientific knowledge—a fortiori the idea of economic knowledge—has changed, as has the very idea of a rigorous scientific argument because of the emergence of the axiomatic approach in mathematics. Thus we have the split, looking ahead to today from the early decades of this century, between those who would argue that mathematical rigor (and scientific knowledge) must develop not from axioms but from observations (about the economy) and (economic) data, so that the truth of a theory or model may be tested or confirmed by reality—like Volterra, Pareto, and Edgeworth—and those who would claim that mathematical (economic) models are rigorous (and "true" in the only useful scientific sense of the word) if they are built on a cogent axiom base—like von Neumann and Morgenstern, and Debreu. The arguments about formalism in economics thus recapitulate divergent views about, and changing meanings of, scientific knowledge. Our archaeology of formalism in economics unearths increasingly energetic and successful challenges to certain more or less traditional or standard views about scientific truth/knowledge, and the development of more or less successful alternatives in various quarters:[15] the strata are the emergent reconceptualizations of both science and knowledge. In concrete terms, there is indeed a disjunction between Debreu's *The Theory of Value* and Friedman and Schwartz's *A Monetary History of the United States, 1867–1960:* although both are mathematically rigorous, the latter is rigorous in an older sense, the former in the newer sense. This is one source of the divergence between econometrics and mathematical economics.

4   Bourbaki and Debreu

One could say that the axiomatic method is nothing but the "Taylor system" for mathematics.—N. Bourbaki, "The Architecture of Mathematics"

Economists, methodologists, and historians of economics have debated the impact and significance of the substantial racheting upward of standards of mathematical sophistication within the profession. Nevertheless, these methodological disputes have been prosecuted in a profoundly ahistorical and internalist fashion. This chapter approaches such issues obliquely, by examining how one distinctive image of mathematics could make inroads into a seemingly distant field and subsequently transform not only that field's self-image, but its conception of inquiry itself. Specifically we will look at how the Bourbakist school of mathematics rapidly migrated into neoclassical mathematical economics in the postwar period, and tell that story primarily through an intellectual biography of a single actor, the Nobel Prize winner Gerard Debreu.

Purity and Danger

Why should the story of the activities of one economist be significant? The answer lies in the way it illustrates the intersection of technical, philosophical, and historical concerns. It describes what happens when the sublimity of pure mathematics (the "music of reason," as the Bourbakist Dieudonne calls it) meets the impurity of scientific discourse, here economics. Too often, such issues are treated merely as matters for the odd speculation about reasons for the "unreasonable effectiveness" of mathematics in the sciences. But as any reader of Mary Douglas can attest, reflection on the impure involves reflection on the relationship of order to disorder: "Rituals of purity and impurity create unity in experience. By their means, symbolic patterns are worked out and publicly displayed. Within these patterns disparate elements are related and disparate experience is given meaning" (Douglas 1989, 2–3). For our purposes, the school of Bourbaki will serve to represent the champions of purity within the house of twentieth-century mathematics.

While Bourbaki is hardly a household word amongst historians, many mathematicians would agree that

> for a few decades, in the late thirties, forties and early fifties, the predominant view in American mathematical circles was the same as Bourbaki's: mathematics is an autonomous abstract subject, with no need of any input from the real world, with its own criteria of depth and beauty, and with an internal compass for guiding future growth. . . . Most of the creators of modern mathematics—certainly Gauss, Riemann, Poincaré, Hilbert, Hadamard, Birkhoff, Weyl, Wiener, von Neumann—would have regarded this view as utterly wrongheaded. (Lax 1989, 455–56)

> The twentieth century has been, until recently, an era of "modern mathematics" in a sense quite parallel to "modern art" or "modern architecture" or "modern music." That is to say, it turned to an analysis of abstraction, it glorified purity and tried to simplify its results until the roots of each idea were manifest. These trends started in the work of Hilbert in Germany, were greatly extended in France by the secret mathematical club known as "Bourbaki," and found fertile soil in Texas, in the topological school of R. L. Moore. Eventually, they conquered essentially the entire world of mathematics, even trying to breach the walls of high school in the disastrous episode of the "new math." (Mumford 1991)

Thus Bourbaki came to uphold the primacy of the pure over the applied, the rigorous over the intuitive, the essential over the frivolous, the fundamental over what one member of Bourbaki called "axiomatic trash." They also came to define the disciplinary isolation of the mathematics department in postwar America. It is this reputation for purity and isolation that drew the wrath of many natural scientists in the 1990s. For instance, the physicist Murray Gell-Mann has written, "The apparent divergence of pure mathematics from science was partly an illusion produced by the obscurantist, ultra-rigorous language used by mathematicians, especially those of a Bourbakist persuasion, and by their reluctance to write up non-trivial examples in explicit detail. . . . Pure mathematics and science are finally being reunited and, mercifully, the Bourbaki plague is dying out" (1992, 7). Or one might cite the case of Benoit Mandelbrot, all the more poignant because of his blood relation to a member of Bourbaki:

> The study of chaos and fractals ought to provoke a discussion of the profound differences that exist . . . between the top down approach to knowledge and the various "bottom up" or self-organizing approaches. The former tend to be built around one key principle or structure, that is, around a tool. And they rightly feel free to modify, narrow down, and clean up their own scope by excluding everything that fails to fit. The latter tend to organize themselves around a class of problems. . . . The top down approach becomes typical of most parts of mathematics, after they have become mature and fully self-referential, and it finds its over-fulfillment and destructive caricature in Bourbaki. The serious issues were intellectual strategy, in mathematics and beyond, and raw political power. An obvious manifestation of intellectual strategy concerns "taste." For Bourbaki, the fields to encourage were few in number, and the fields to discourage or suppress were many. They went so far as to exclude (in fact, though perhaps not in law) most of hard classical analysis. Also unworthy was most of sloppy science, including nearly everything of future relevance to chaos and to fractals. (Mandelbrot 1989, 10–11)

For many scientists, Bourbaki became the watchword for the chasm that had opened up between mathematics and its applications, between the rigor of axiomatization and rigor in the older sense (see chapter 2) of basing argumentation on the physical problem situation. In such a world, would it not appear that a Bourbakist-inspired discipline of "applied mathematics" would be an oxymoron? It is our thesis that such a thing did occur in economics, and indeed, it took root and flourished in the postwar American environment. The transoceanic gemmule was Gerard Debreu; the seedbed for economics was the Cowles Commission (Christ 1952) at the University of Chicago. This particular narrative demonstrates just how the pure and the impure were constantly intermingled in mathematical practice, suggests some of the attractions and dangers that fertilized the transplant, and perhaps also opens up the hothouse of mathematics to a historiographic search for the influence of Bourbaki and other such versions of "images of mathematics" (Corry 1989) upon the whole range of the sciences in the twentieth century.

Pure Structures for an Impure World

Who or what was Bourbaki, that they could so utterly transform the staid world of mathematics? While primary materials are sparse, and no comprehensive history in English exists, we shall base our brief narrative on the published texts by Bourbaki, some statements about Bourbaki by former members (Dieudonne 1970, 1982; Cartan 1980; Guedj 1985; Adler 1988) and the important papers by Corry (1992a, 1992b). Our intention is primarily to set the stage for the appearance of our protagonist, Gerard Debreu, and not to provide anything like a comprehensive overview of the Bourbaki phenomenon.

In 1934–35, Claude Chevalley and Andre Weil decided to try to reintroduce rigor into the teaching of calculus in France by rewriting one of the classic French treatises. As Chevalley recalled matters, "The project, at that time, was extremely naive: the basis for teaching the differential calculus was Goursat's *Traite*, very insufficient on a number of points. The idea was to write another to replace it. This, we thought, would be a matter of one or two years" (Guedj 1985, 19). The project (which continues to this day) was adopted as the work of the original group of seven; in the Bourbaki nomenclature they are called the "founders": Henri Cartan, Claude Chevalley, Jean Delsarte, Jean Dieudonne, Szolem Mandelbrojt, Rene de Possel, and Andre Weil. Continuing an elaborate joke that had been played, over time, at the École Normale Superieur, they gave themselves the name of an obscure nineteenth-century French general, Nicolas Bourbaki, and agreed to operate as a secret club or society.[1] At the beginning, they also agreed that the model for the book they wished to write was B. L. Van der Waerden's *Algebra*, which had appeared in German in 1930. "So we intended to do something of this kind. Now Van der Waerden uses very precise language and has an extremely tight organization of the development of ideas and of the different parts of the work as a whole. As this seemed to us to be the best way of setting out the book, we had to draft many things which had never before been dealt with in detail" (Dieudonne 1984 [1970], 106).

The difficulty was that this project was an immense one, and "we quickly realized that we had rushed into an enterprise which was considerably more vast than we had imagined" (ibid.). The work was done in occasional meetings in Paris, but mostly in "congresses," the longest of which took place in the French countryside each summer. The rules of Bourbaki quickly became established, both the formal and informal ones. Of the formal rules, there was only one, and that was that no member of the group should be over age fifty, and that on reaching that age, a member would give up his place. Nevertheless, certain behaviors became conventional. There came to be two meetings a year in addition to the longer congress. The work was done by individuals agreeing to submit drafts of chapters to the group for public reading, and for tearing apart. If the result was not accepted—and acceptance required unanimity—then the draft was given to someone else to be rewritten and resubmitted at a subsequent congress. Up to two visitors might attend the congresses, provided they participated fully; this was sometimes a way to see if a person might be thought of as a potential new Bourbaki.

> There never was an example of a first draft being accepted. The decisions did not take place in a block. In the Bourbaki congresses one read the drafts. At each line there were suggestions, proposals for change written on a black-board. In this way a new version was not born out of a simple rejection of a text, but rather it emerged from a series of sufficiently important improvements that were proposed collectively. (Guedj 1985, 20)

The question of what kind of book they were to write quickly came to the forefront of their discussions. What distinguishes Bourbaki's project is the result of the Bourbaki decision to create a "basic" book for mathematicians.

The idea which soon became dominant was that the work had to be primarily a tool. It had to be something usable not only in a small part of mathematics, but also in the greatest possible number of mathematical places. So if you like, it had to concentrate on the basic mathematical ideas and essential research. It had to reject completely anything secondary that had immediately known application [in mathematics] and that did not lead directly to conceptions of known and proved importance. . . . So how do we choose these fundamental theorems? Well, this is where the new idea came in: that of mathematical structure. I do not say it was a new idea of Bourbaki—there is no question of Bourbaki containing anything original. . . . Since Hilbert and Dedekind, we have known very well that large parts of mathematics can develop logically and fruitfully from a small number of well-chosen axioms. That is to say, given the bases of a theory in an axiomatic form, we can develop the whole theory in a more comprehensible form than we could otherwise. This is what gave the general idea of mathematical structure. . . . Once this idea had been clarified, we had to decide which were the most important mathematical structures. (Dieudonne 1984 [1970], 107)

By 1939 the first book appeared, *Elements de Mathematique, Livre I (Fascicule de resultats)*. This book was the first part of the first volume, that of set theory. It presented the plan of the work, and outlined the connections between the various major parts of mathematics in a functional way, or what Bourbaki called a structural manner. It contained

without any proof all notations and formulas in set theory to be used in subsequent volumes. Now when each new volume appears, it takes its logical position in the whole of the work. . . . Bourbaki often places an historical report at the end of a chapter. . . . There are never any historical references in the text itself, for Bourbaki never allowed the slightest deviation from the logical organization of the work itself. (Cartan 1980 [1958], 8)

Thus, instead of the division into algebra, analysis, and geometry, the fundamental subjects from which the others could be derived, were to be set theory, general algebra, general topology, classical analysis, topological vector spaces, and integration. This organization shows up in the volumes themselves, for the first six books, each of several chapters with numerous exercises, correspond to these six divisions. The twenty-one volumes published by the late 1950s all belong to part I, "The Fundamental Structures of Analysis."

"An average of 8–12 years is necessary from the first moment we set to work on a chapter to the moment it appears in a bookshop" (Dieudonne 1984 [1970], 110). The length of time seems to be a result of both the unanimity rule for the congresses, and the complexity of the task itself. "What was envisioned was a repertory of the most useful definitions and theorems (with complete proofs . . .) which research mathematicians might need . . . presented with a generality suited to the widest possible range of applications. . . . In other words, Bourbaki's treatise was planned as a bag of tools, a tool kit, for the working mathematician" (Dieudonne 1982, 618).

This viewpoint led to the fundamental organizing idea of the work: "It was our purpose to produce the general theory first before passing to applications, according to the principle we had adopted of going 'from the general (*generalissime*) to the particular' " (Guedj 1985, 20). Of the views of the import of the project, the "founders" seemed to believe, as Chevalley recalled, "It seemed very clear that no one was obliged to read Bourbaki . . . a bible in mathematics is not like a bible in other subjects. It's a very well arranged cemetery with a beautiful array of tombstones. . . . There was something which oppressed us all: everything we wrote would be useless for teaching" (Guedj 1985, 20).

It was to be through the Seminaire Bourbaki that the French mathematicians reconnected to the world mathematical community after World War II. The project of the Elements gained momentum, and the invitations to come to lecture in Paris were appreciated. The immense intellectual strength of the French mathematicians in a number of important areas made it more and more noted among mathematicians in the United States. The international nature of the mathematical community, and the prewar connections of the few older men, Andre Weil particularly, facilitated recognition of the work. The mystery of Bourbaki himself, and the ambition of his project, probably encouraged attention as well.

Bourbaki had the major problem, in writing the Elements, of organization, of relating the various parts of mathematics one to another. This "problem" was approached through the notion of "mathematical structure," of which more anon. The second issue Bourbaki had to face was that of the approach to be taken within each section of the whole, and that was handled by the rule "from the general to the specific." Thus as the books and chapters emerged from the publisher, and the immense project took shape in print over the decades, mathematics was presented as self-contained in the sense that it grew out of itself, from the basic structures to those more derivative, from the "mother-structures" to those of the specific areas of mathematics. For example,

> In the logical order of the Bourbaki system, real numbers could not appear at the beginning of the work. They appear instead in the fourth chapter of the third book. And with good reason, for underlying the theory of real numbers is the simultaneous interaction of three types of structures. Since Bourbaki's method of deducing special cases from the most general one, the construction of real numbers from the rationals is for him a special case of a more general construction: the completion of a topological group (Chapter 3 in Book III.) And this completion is itself based on the theory of the completion of a "uniform" space (Chapter 2 in Book III). (Cartan 1980 [1958], 178)

What these organizing principles accomplished, in making the work itself coherent, cannot be underestimated. The choices Bourbaki made were reasonable ones for the immense task of writing a handbook of mathematics for working mathematicians. The imposition of this order, and coherence, led to a book with the elegance and grace of a masterwork, a modern version of Euclid's *Elements of Geometry*. But the ideas of structure, and the book's movement from the general to the specific, had major consequences.

The word "structure," whether in French or in English, can mean many things to many people. The immediate temptation is to associate it with the erstwhile French philosophical and cultural movement known as "structuralism" (Caws 1988); there is some justification for this inclination, such as the connections between Andre Weil and one of the gurus of the movement, Claude Lévi-Strauss. Indeed, the title of one of Bourbaki's very few explicitly methodological pronouncements, published in 1948, was "The

Architecture of Mathematics," anticipating the title and some of the content of Michel Foucault's own *L'archeologie du Savoir* by two decades (Gutting 1989). We shall regretfully bypass such tantalizing historical issues and opt to concentrate more narrowly upon Bourbaki's own account of the meaning of structure, and the clarification of these issues provided by Corry (1992a).[2]

The question that motivated Bourbaki was, "Do we have today a mathematics or do we have several mathematics?" (1948, 221). Fears of disorder, or "dirt" as Mary Douglas would put it, were the order of the day, with Bourbaki (1950, 221) wondering, "Whether the domain of mathematics is not becoming a Tower of Babel?" Bourbaki would not want to pose this question to the philosophers, but rather to an ideal type that he identified as the "working mathematician." This *l'homme moyen* was purportedly defined by his recourse to "mathematical formalism": "The method of reasoning by laying down chains of syllogisms. . . . To lay down the rules of this language, to set up its vocabulary and to clarify its syntax" (223). Bourbaki goes on to state, however, that this is

> but one aspect of the axiomatic method . . . [which] sets as its essential aim . . . the profound intelligibility of mathematics. . . . Where the superficial observer sees only two, or several, quite distinct theories . . . the axiomatic method teaches us to look for the deep-lying reasons for such a discovery, to find the common ideas of these theories, buried under the accumulation of details properly belonging to each of them, to bring these ideas forward and to put them in the proper light. (223)

He then proceeds to suggest that the starting point of the axiomatic method is a concern with "structures," and develops this idea of structure through examples. The informal definition is that a structure is a

> generic name . . . [which] can be applied to sets of elements whose nature has not been specified; to define a structure, one takes as given one or several relations, into which these elements enter (in the case of groups, this was the relation $z = x\tau y$ between the three arbitrary elements); then one postulates that the given relation, or relations, satisfy certain conditions (which are explicitly stated and which are the axioms of the structure under consideration. To set up the axiomatic

theory of a given structure, amounts to the deduction of the logical consequences of the axioms of the structure, excluding every other hypothesis on the elements under consideration (in particular, every hypothesis as to their own nature). (225–26)

This remarkable passage is in fact the linchpin of the enterprise, for it contains in it, and outside it by what it excludes, Bourbaki mathematics.

First, note "nature" is footnoted. Bourbaki comments that philosophical concerns are to be avoided here, in the debates on formalist, idealist, intuitionist foundations. Instead, "from this new point of view mathematical structures become, properly speaking, the only 'objects' of mathematics" (225n–26n). That is, mathematics is concerned with mathematical objects, called structures, if you will, and the job of mathematicians is to do mathematics attending to these structures. Bourbaki goes on to say, in a footnote to the word "enter," that "this definition of structures is not sufficiently general for the needs of mathematics" because of a need to consider higher order structures, or in effect structures whose elements are structures. The Gödel incompleteness issues are left to one side, for mathematicians simply do mathematics, and when an inconsistency arises, the rule is to face it, and do mathematics around it almost in an empirical sense.

> What all this means is that mathematics has less than ever been reduced to a purely mechanical game of isolated formulas; more than ever does intuition dominate in the genius of discoveries. But henceforth, it possesses the powerful tools furnished by the great types of structures; in a single view, it sweeps over immense domains, now unified by the axiomatic method, but which formerly were in a completely chaotic state. (228)

In the 1949 paper Bourbaki lays out his actual program for foundations in the post-Gödel world of logic:

> What will be the working mathematician's attitude when confronted with such [Gödel] dilemmas? It need not, I believe, be other than strictly empirical. We cannot hope to prove that every definition, every symbol, every abbreviation that we introduce is free of potential ambiguities, that it does not bring about the possibility of a contradiction that might not otherwise have been present. Let the rules be so formulated, the definitions so laid out, that every contradiction may be most

easily traced back to its cause, and the latter either removed or so surrounded by warning signs as to prevent trouble. This, to the mathematician, ought to be sufficient; and it is with this comparatively modest and limited objective in view that I have sought to lay the foundations for my mathematical treatise. (Bourbaki 1949, 3)

What we have here is the "admission" that there is no more security to be found in the magisterial idea of "structure" than there was in the idea of "set" or "number" as the bedrock on which a secure mathematics could be built. Nonetheless, Bourbaki lays out, in this paper, the "sign-language" of objects, signs, relations to end up with a language in which he can proceed to do mathematics. That this is not necessarily consistent is of no concern to the working mathematician, for it suffices to do the Bourbaki mathematics. We have then a disjunction between what Corry calls "structure" and *structure*.

Leo Corry (1989) suggested that mathematics should be set apart from other sciences because it persistently strives to apply the tools and criteria of its actual practices to itself in a meta-analytic manner, thus masking the distinction between the "body of knowledge" characteristic of a particular historical juncture and the "images of knowledge" that are deployed in order to organize and motivate inquiry. For Corry, it is the images of knowledge rather than the actual corpus of proofs and refutations that gets overthrown and transformed whenever mathematical schools and fashions change over the course of history. Corry's premier illustration of this thesis is his (1992a) description of Bourbaki's variant meanings of the terms "structure" and *structure*.

In the 1939 Fascicule, hereafter cited as the Theory of Sets, Bourbaki proposed to lay out the foundations in chapter 4, the formally rigorous basis of their entire enterprise. This collection of formalisms, which Corry designates as *structure*, involved base sets and an echelon construction scheme that was intended to generate mother-structures, which in turn would generate the rest of mathematics as Bourbaki saw it. Yet there was a disjuncture between this chapter and the rest of the book, as well as with all the other volumes of the Bourbaki corpus. "Bourbaki's purported aim in introducing such concepts is expanding the conceptual apparatus upon which the unified development of mathematical theories would rest later on. However all this work turns out to be rather redundant since . . . these concepts are used in a very limited—and certainly not highly illuminating or unifying—fashion in the remainder of the treatise" (Corry 1992a, 324). It seems the concept of structure has no palpable mathematical use in the rest of Bourbaki's work, and the links between the formal apparatus and the working mathematician are largely absent. "No new theorem is obtained through the structural approach and standard theorems are treated in the standard ways" (1992a, 329). Yet, as we have already witnessed, the ideal of "structure" and the achievement of Bourbaki have remained identified in the minds of those who came afterward. How can this be?

Corry responds that this has to do with the difference between the actual body of results and the image of knowledge. "If the book's stated aim was to show that we can formally establish a sound basis for mathematics, the Fascicule's purpose is to inform us of the lexicon we will use in what follows and of the informal meaning of the terms within it. The sudden change in approach, from a strictly formal to a completely informal style, is clearly admitted" (1992a, 326). This is the practical meaning of Corry's unitalicized term "structure": Bourbaki's primary contribution had to do with the way mathematicians interpreted their mathematical work, and not the formal foundations of that work itself. It was, if you will, a matter of style, of taste, of shared opinions about what was valuable in mathematics, of all those things that shouldn't really matter to the Platonist or the Formalist or the Intuitionist. Or as Corry put it, "Bourbaki's style is usually described as one of uncompromising rigor with no heuristic or didactic concessions to the reader. . . . [But in The Theory of Sets] the formal language that was introduced step by step in Chapter I is almost abandoned and quickly replaced by the natural language" (1992a, 321).

The final legacy of Bourbaki is most curious. As Corry summarized (1992b, 15), "Bourbaki did not adopt formalism with full philosophical commitment, but rather as a facade to avoid philosophical difficulties." Others now concur in this assessment (Mathias 1992). Bourbaki gave the impression of elevating his choices in mathematics above all dispute: but that was all it was—just an impression.

> It is clear that the early developments of the categorical formation, more flexible and effective than the one provided by structures rendered questionable Bourbaki's initial hopes of finding the single best foundation for each mathematical idea and cast doubt on the initially

intended universality of Bourbaki's enterprise. . . . [As Saunders Mac Lane wrote], good general theory does not search for the maximum generality, but for the right generality. (Corry 1992a, 336)

But this realization took time, happening possibly as late as the 1970s; and in the interim, the Bourbaki juggernaut kept churning out further volumes. The timing of these events will be of some significance for our subsequent narrative.

These details concerning Bourbaki's history and Corry's reading of it, seemingly so far removed from economics, are instead absolutely central to understanding its postwar evolution. The reason is that very nearly everything said about Bourbaki will apply with equal force to Gerard Debreu.

Gerard Debreu and the Making of a Pure Economics

When the place of mathematics in twentieth-century economics is broached, it is Debreu who is always mentioned with awe, and not a little apprehension. "Debreu is known for his unpretentious no-nonsense approach to the subject," writes Samuelson (1983, 988). "Debreu's contributions might appear, at first glance, incomprehensibly 'abstract.' . . . In this respect Debreu has never compromised just as he has never followed fashions in economic research," writes his memorialist Werner Hildenbrand. "Debreu presents his scientific contributions in the most honest way possible by explicitly stating all underlying assumptions and refraining at any stage of the analysis from flowery interpretations that might divert attention from the restrictiveness of the assumptions and lead the reader to draw false conclusions" (Hildenbrand 1983, 2–3). When George Feiwel tried to conduct an oral history, he was reduced to prefacing many of his questions with the clause, "For the benefit of the uneducated." In response to the question, "Why is the question of existence of general economic equilibrium so profoundly important?," Debreu shot back, "Since I have not seen your question discussed in the terms I would like to use, I will not give you a concise answer" (Feiwel 1987, 243). However, when I interviewed him in 1992, he was gracious and forthcoming in answering many questions about his career (see later in this chapter for the interview in its entirety).

Debreu is perhaps best known for his 1954 joint proof with Kenneth Arrow of the existence of a general competitive Walrasian equilibrium (Weintraub 1985) and his 1959 monograph *The Theory of Value*, which still stands as the benchmark axiomatization of the Walrasian general equilibrium model. In retrospect, the 1959 book wore its Bourbakist credentials on its sleeve, though there may have been few economists at that juncture who would have understood the implications of this statement:

> The theory of value is treated here with the standards of rigor of the contemporary formalist school of mathematics. The effort toward rigor substitutes correct reasonings and results for incorrect ones, but it offers other rewards too. It usually leads to a deeper understanding of the problems to which it is applied, and this has not failed to happen in the present case. It may also lead to a radical change of mathematical tools. In the area under discussion it has been essentially a change from the calculus to convexity and topological properties, a transformation which has resulted in notable gains in the generality and the simplicity of the theory. Allegiance to rigor dictates the axiomatic form of the analysis where the theory, in the strict sense, is logically entirely disconnected from its interpretations. In order to bring out fully this disconnectedness, all the definitions, all the hypotheses, and the main results of the theory, in the strict sense, are distinguished by italics; moreover, the transition from the informal discussion of interpretations to the formal construction of the theory is often marked by one of the expressions: "in the language of the theory," "for the sake of the theory," "formally." Such a dichotomy reveals all the assumptions and the logical structure of the analysis. (Debreu 1959, x)

While it was the case that most economists would have been unfamiliar at that time with the novel tools of set theory, fixed point theorems, and partial preorderings, there was something else that would have taken them by surprise: a certain take-no-prisoners attitude when it came to specifying the "economic" content of the exercise. Although there had been quantum leaps of mathematical sophistication before in the history of economics, there had never been anything like this. For instance, few would have readily recognized the portrait of an "economy" sketched in the monograph:

An economy E is defined by: for each i=1, . . . m a non-empty subset x^i of R^l completely preordered by $\leq_i$ [at least as desirable to agent i]; for each j=1, . . . n a non-empty subset of Yj of R^l; a point ω of R^l. A state of E is an (m+n)-tuple of points of R^l. (Debreu 1959, 75)

While more than one member of the profession might have thought this species of economist had dropped from Mars, in fact he had merely migrated from France. The way that this happened might go some distance in explaining the otherwise totally unprecedented character of this kind of mathematical economics.

Gerard Debreu was born on 4 July 1921 in Calais, France. He experienced a successful early school career preparing for the Baccalaureate by studying physics and mathematics. His plans to study at a Lyceé for entrance into one of the Grandes Écoles were disrupted by the beginning of the war, but he did manage further preparation in mathematics at Grenoble; he won the Concours General in 1939 in physics, and later admission into the École Normale Supérieure.

The group entering the École Normale Supérieure was divided roughly in half, with around fifty students each, in humanities and sciences. Around twenty of the fifty science students were mathematics students.

The sciences were divided basically between mathematics on one hand and physics and chemistry on the other (the two went together) and there was a third possibility (but very few students went that way), that was biology. And I imagine that in our group maybe only one or two went the way of biology whereas the division between mathematics and physics and chemistry was about even. . . . All science students took the same examination [to] enter the school, and then we decided which way to go. In mathematics it was normally a three-year course and in physics I think it was four. (131–32)

The mathematical training that Debreu received at the École Normale Supérieure was very different from that which he had had earlier. Instruction was carried out, in mathematics, in a complicated fashion.

It's very strange. Again, it is unique. If you take another Grand École like the École Polytechnique, they have all their teaching within the school only for students there. Not at the École Normale Supérieure. It is close to the Sorbonne, geographically close, and we were supposed to take the standard courses at the Sorbonne. And what we had at the École Normale Supérieure was very small seminars; that is where we were taught by Cartan. There was no fixed curriculum, and it was attended by about 10 people whereas in the fundamental courses at the Sorbonne the attendance was at first in the hundreds. . . . What was lacking in them then was the enthusiasm that Cartan generated. (Ibid.)

For Debreu, the mathematical work was interesting, but he already had some idea that he was perhaps going to be more involved with mathematics in another discipline. Perhaps this was because of his earlier success in physics, perhaps it is because he reached a limit in his ability to sustain interest in pure mathematics under the conditions of wartime Paris. At any rate, Debreu seems to have understood, fairly early in his career at the École Normale Supérieure, that his own path was to be a bit different from those of his fellow mathematicians.

The objective at the École Normale Supérieure was basically to produce teachers of mathematics; and that was understood in the days when I was there, to mean teachers of mathematics at the Mathématiques Spéciale Préparatoire and Mathématiques Spéciale level. . . . After a year or so (I entered in the fall of 1941), I began to wonder whether mathematics at that time was becoming too very abstract under the influence of Bourbaki. . . . I had to decide whether I wanted to spend my entire life doing research in a very abstract subject. You must also remember that during that last year of high school when I was influenced by my physics teacher I had thought that physics was going to be my field. (Ibid.)

His training at the École Normale Supérieure was at the highest university level, and in fact can better be compared to the work done at the graduate level at most other universities, because they had to do the standard university mathematics curriculum on their own. The salient point is that Debreu was as well trained in mathematics as was possible for any student to have been at that time. He had the remarkable fortune to be at the place, at the time, when mathematics itself was being re-represented by Bourbaki as a discipline defined by its pursuit of the implications of, and the investigation and exposition of, the idea of mathematical structure. In this mathematical hothouse, isolated because of the war and the dislocations it produced, Debreu pursued mathematics but did not want to have it define his intellectual life. But there were no real alternatives; he was a mathematics student first, and other possibilities would have to be deferred to the end of the war, since the only applied alternative that he might have considered at École Normale Supérieure, astrophysics, seemed to be ruled out by the absence of any real instructional program—the professor was not present. Stuck in mathematics, around 1943 he looked at possibilities for later work:

> When I became interested in economics as a possibility (as before I had become interested in astrophysics) I got hold of the standard text studied by students of economics at the university. I don't remember who the author was but it was very non-theoretical (somebody I never met). I know the textbook was popular then—I don't believe I have kept a copy—but in any case, my first impression of economics was very disappointing because I was coming from a world of very sophisticated and rarified mathematics and found only a very pedestrian approach to economics. (Ibid.)

Debreu has recounted the happenstance of his receiving a copy of Allais's 1943 book, *A la Recherche d'une Discipline Économique*. The book arrived in Debreu's hands at a very crucial moment, for Debreu was searching for meaningful work, as many young people search at that age. The Allais book had been more or less just sent around to individuals; it had been essentially privately printed. It was very much outside the established French economics channels. In retrospect, it is remarkable that it was able to be printed at all under those wartime conditions, but also that it received any hearing given the number of unusual books that drop out of sight. Even had it gone to all those interested in mathematical economics at that time, there would have been problems. It was very primitive mathematics from a Bourbaki point of view, though it was more sophisticated than most neoclassical texts. Nevertheless, in Debreu it found a friendly reader. "First of all, I saw that mathematics could be used in economics in a rigorous way, even though it was not the kind of mathematics I was most fond of. And maybe I felt that there was a lot to be done with more sophisticated mathematics in economics. My interest in economics wasn't ready made" (ibid.).

The story, in outline, is clear at this point. Debreu was a very well-trained research mathematician in the Bourbaki mold. However, he also fits a rather common profile of many key figures in political economy from the 1930s and 1940s: someone with very little background in economics moving over into that field following a thorough academic training in physics or mathematics (Mirowski 1991). The engineer-autodidact Allais was not at all representative of the state of political economy in France at the time, and as a consequence Debreu had to find his own way around the mathematical economics literature; he was, by his own account, particularly impressed by John von Neumann and Oskar Morgenstern's *Theory of Games and Economic Behavior* (1944) in this period. From 1946 to 1948 he occuied the position of Research Associate at the Centre National de la Recherche Scientifique; upon being awarded a Rockefeller Foundation travel fellowship, he toured Harvard, Berkeley, Chicago, Uppsala, and Oslo: the most fateful of those visits was the sojourn at the Cowles Commission at the University of Chicago.

His appearance at Cowles in 1949 was fortuitous. The Cowles Commission up until that point had been primarily known as a center for the development of econometrics—the application of mathematical statistics to empirical economic questions—but various crises having to do with disappointments in their program of structural estimation and turf battles with the Economics Department at Chicago was causing the unit to contemplate a change in research direction (Epstein 1987, 110; Mirowski 1993). The research director at the time was Tjalling Koopmans, a Dutch refugee from quantum physics whose prior work had primarily involved statistical estimation. The reorientation of research away from empirical work and toward mathematical theory had already begun under Koopmans by 1949, but it lacked clear direction. Debreu felt right at home among the mathematically sophisticated advocates of neoclassical economics, many of them also European expatriates with degrees in the natural sciences. However, there was another, serendipitous, side to Chicago. Spurned by the economists, Koopmans had begun making overtures to the Mathematics Department to establish a mathematical statistics unit. The chair of the department was Marshall Stone, one of the main boosters of Bourbakism in the American context. Stone had been reshaping the department in a Bourbakist direction since 1947, attracting Andre Weil and building a first-class mathematics research faculty (Stone and Browder in

Duren et al. 1989). Koopmans kept in close contact with Stone through the Committee on Statistics.

The exact vectors of influence are unclear, but after Debreu permanently joined the Cowles Commission in June 1950, Bourbakism quickly became the house doctrine of the Cowles Commission. We would identify the primary philosophical texts asserting this turning point as Koopmans's *Three Essays on the State of Economic Science* (1957) and Debreu's *Theory of Value* (1959). The former was the classroom primer of the new approach, with explicit methodological discussions of the nature of mathematical rigor and the relation of economics to practices in physics; whereas the more austere *Theory* was intended to show how cutting-edge research would be done in the future. Debreu explicitly signposted *Three Essays* as facilitating the understanding of his own work (1959, x). While Koopmans and Debreu were the main proponents of this new approach, both subsequently winning Nobel Prizes for their work dating from this era, one can also observe the new orientation in the work of others associated with Cowles in this period: John Chipman, Murray Gerstenhaber, I. N. Herstein, Leonid Hurwicz, Edmond Malinvaud, Roy Radner, and Daniel Waterman. When the Cowles Commission moved to Yale in 1955, the Bourbakist attitudes toward mathematical theory began to spread throughout American graduate education in economic theory, as the "Cowlesmen" fanned out into the major economics departments.

Why the Bourbakist orientation toward mathematical economics spread so rapidly within the American context, once it had been crystallized within the Cowles Commission, is just being addressed in the literature (Mirowski 2001; Leonard forthcoming). But a few generalizations might be suggested here, before we focus directly once more upon Debreu. First, since Cowles had become disillusioned with its earlier empiricist commitments, the Bourbakist program of isolation of theory from its empirical inspiration proved both convenient and timely. Skepticism about the quality of economic empiricism became a hallmark of those who took up the program of mathematical formalization. Moreover, this turn away from empiricism mirrored the turn away from such work as "rigorous" in applied mathematics more generally (as discussed in chapter 2).

Second, it is sometimes forgotten that the 1940s was a period of much rivalry and dissension among diverse schools of economic thought, and that Cowles often found itself in the thick of controversy. For instance,

Cowles itself was battling Wesley Clair Mitchell's Institutionalists at the National Bureau of Economic Research (NBER) for funding and legitimacy in the "Measurement without Theory" debate; Keynesians like Lawrence Klein at Cowles were at odds with Milton Friedman and others in the Chicago economics department. Bourbakism held out the promise of rising above it all, offering a vantage point from which one might stand aloof from the Babel that threatened to drown out reasoned discourse. It almost became a badge of honor to suggest that one was not familiar with previous traditions in economics: for instance, Koopmans appears to have written in a proposal to the Ford Foundation, "With one possible exception (Simon) the present staff of the Commission can claim no special competence in the tomes of social science literature, nor do we think that this should be a primary criterion in the selection of additional staff . . . we intend in staffing to give greatest weight to the combination of creative imagination and rigorous logical and/or mathematical treatment of problems" (unattributed "Application to Ford Foundation," submitted 17 September 1951, p. 14; unboxed materials, Cowles Foundation Archives, Yale University).

And third, one should not forget that Bourbakism was sweeping the American mathematics profession in this same period. Many social sciences made concerted efforts to mathematize their doctrines in the immediate postwar era, but it was only economists who seemed to be doing mathematics of a sort that a mathematician would recognize. Indeed, Debreu was named to a position in the mathematics department at Berkeley in 1975, over and above the Professorship of Economics he had held there since 1962.

The rise of mathematical formalism in economics is not a simple phenomenon of the imperative of the subject matter, as it is sometimes claimed; rather it is the product of contingencies of the intersection of diverse disciplines and, as Debreu is the first to acknowledge, numerous personal accidents and fortuitous encounters.

Setting the Structures Aright

When Debreu was awarded the Nobel Prize in 1983, many reporters and commentators were flummoxed by their encounter with this austere economist. His work was abstruse and impenetrable, his demeanor reserved, and his resistance to using the bully pulpit in order to comment upon current economic events unprecedented. Many within the economics profession have likewise found his program inscrutable, because they insist on trying to frame it in their own local terms. We should like to suggest that better interpretative headway could be made if the analogies to the Bourbakist program in mathematics were taken much more seriously. Indeed, many aspects of the Bourbakist program find direct correspondences in the details of Debreu's version of mathematical economics.

It seems clear that Debreu intended his *Theory of Value* to serve as the direct analogue of Bourbaki's *Theory of Sets,* right down to the title. His monograph sought to establish the definitive analytic mother-structure from which all further work in economics would depart, primarily either by "weakening" its assumptions or else superimposing new "interpretations" upon the existing formalism. But this required one very crucial maneuver that was nowhere stated explicitly: namely, that the model of Walrasian general equilibrium was the root structure from which all further scientific work in economics would eventuate. As perceptively noted by Ingrao and Israel (1990, 286), "In Debreu's interpretation, general equilibrium theory thus loses its status as a 'model' to become a self-sufficient formal structure." The objective was no longer to represent the economy, whatever that might mean, but rather to codify the very essence of that elusive entity, the Walrasian system. This fundamental shift in objective explains many otherwise puzzling features of Debreu's career, such as the progressive shift away from his early dependence upon game theoretic concepts, his disdain for attempts (like that of Kenneth Arrow and Frank Hahn) to forge explicit links between the Walrasian model and contemporary theoretical concerns in macroeconomics or welfare theory, and his self-denying ordinance in dealing with issues of stability and dynamics.

Just as with Bourbaki, the problem was to justify the initial identification of the structures. In Debreu's case, one must insist that this was not a foregone conclusion: Walrasian theory was not widely respected in either France or America; there were alternative versions of the neoclassical program (see Mirowski 2001), like the Marshallian apparatus of demand and supply, with more substantial adherents in America in that era; there existed some rivals to the neoclassical orthodoxy, like Marxism and Institutionalism; and it was only with Joseph Schumpeter's *History of Economic Analysis* (1954) that Walras was identified as "the greatest economist of all time." To believe that the structure of all analytical economics lay half-obscured in the relatively dormant Walrasian/Paretian variant in 1950 was a bold leap of faith. One consideration that may have rendered the leap less unlikely was the fact that Walras presented his own work as the *Elements of Pure Economics*. Here we find a well-articulated notion of the separation of pure theory from its applied aspect; this was sure to resonate with Debreu's inclination to accept a separation of pure from applied mathematics. But another factor, operant for Allais and many of the members of the Cowles Commission, was the similarity of the Walrasian mathematics to structures used in physics (Mirowski 1989).

The importance of the analogy between extrema of field theories in physics and constrained optimization of utility in neoclassical economics was acknowledged on a number of occasions by Koopmans (Mirowski 1991). Since many of the expatriates had little background in economics, the similarities in mathematics initially served to expedite their migrations into the field. Yet the analogy could cut two ways, in that unlike the cases of such individuals as Edgeworth and Jevons, no one in the twentieth century wanted to maintain that utility and energy were ontologically identical. This left the Walrasian program bereft of an explanation of the similarities with physics. Cowles developed an interesting response to this conundrum, namely, that the novel mathematical techniques imported by Koopmans, Debreu, and others liberated economics from its dependence upon classical calculus and physical analogies. Debreu, as noted, took this position even further by claiming that his Bourbakist program marked the definitive break with physical metaphors, since physics was dependent for its success upon bold conjectures and experimental refutations, but economics had nothing else to fall back upon but mathematical rigor. This is entirely consistent with the Bourbakist creed, which acknowledges that mathematical inspiration may originate in the special sciences, but that once the analytical structure is extracted the conditions of its genesis are irrelevant.

In sum, the format of both books mirror each other, with *Theory of Value* exemplifying the ideal of uncompromising rigor, devoid of all heuristic or didactic concessions to the reader. Just as Bourbaki was interested in, and regarded his project as providing a handbook for, the working mathematician, Debreu is best read as providing a handbook for the working economic theorist of the neoclassical components of economic theory. In retrospect, it is hard to read *Theory of Value* as anything else, since it also provides no "new" theorems or results; it is Chevalley's "very well arranged cemetery with a beautiful array of tombstones" (Guedj 1985, 20). Debreu's evident enthusiasm in chapter 7 over his capacity to incorporate "uncertainty" into the axiomatized model by keeping the identical mathematical formalisms but redefining the "interpretation" of the commodity thus should not be regarded as a new contribution to the economic theory of risk or ignorance; rather, in this reading, Debreu developed it as ratification of the structural character of his axioms. Nevertheless, in a manner undoubtedly not intended by Debreu, the monograph also shares many of the same problems of *structures* and "structures" experienced by Bourbaki.

The issue is multilayered but essentially similar at each level. Bourbaki had claimed that the fundamental structures all shared some analytical unifying characteristics; but that claim was asserted, not defended directly: indeed, the book project itself was to be the justification of the assertion. The young Debreu appeared to argue that the Walrasian general equilibrium theory should be treated as possessing the same privileged structural status in economics as "groups" have among "algebraic structures" and the order relation has among "topological structures." But this assertion was ultimately problematized by both Debreu himself and the newer generation of mathematical economists trained to Debreu's standards of rigor— we here refer to what are often cited as the "Sonnenschein/Mantel/ Debreu" results, the importance of which were rendered general currency in the 1980s. But of course, in both cases the set of practices had by that late date gathered its own momentum, to such an extent that both Bourbakist and Debreuvian formalism had come to represent a style of mathematical expression long after they had dropped the role of providing philosophical grounding for their respective disciplinary programs.

And then there was the simple issue of phase lag between the disciplines of mathematics and economics: the disillusion with Bourbaki was evident by the 1970s in mathematics; a similar soul-searching is only now coming to economics. When Debreu first read the Fascicule in the 1940s he had no way of knowing how the Bourbakist structural program would turn out in the 1960s. This perhaps helps explain the rather reserved tone of Debreu's later pronouncements on the place of mathematics in economics,

Before the contemporary period of the last five decades, theoretical physics had been an inaccessible ideal towards which economic theory sometimes strove. During that period, this striving became a powerful stimulus in the mathematicization of economic theory. The great theories of physics cover an immense range of phenomena with supreme economy of expression. This extreme conciseness is made possible by the privileged relationship that developed over several centuries between physics and mathematics. The benefits of that special relationship were large for both fields; but physics did not completely surrender to the embrace of mathematics and to its inherent compulsion towards logical rigor. . . . In these directions, economic theory could not follow the role model offered by physical theory. Being denied a sufficiently secure experimental base, economic theory has to adhere to the rules of logical discourse and must renounce the facility of internal inconsistency. (Debreu 1991, 17)

Debreu, as noted above, has never seemed very interested in describing the dynamics of convergence of an economy to Walrasian equilibrium. The issue of motion could not be avoided forever, however, and there was a long interval in the postwar period in which "dynamics" were redefined to mean "stability" within the mathematical economics community (Weintraub 1991). In that context, the question was posed by Hugo Sonneschein whether the basic "structure" of Walrasian general equilibrium models placed any substantial restrictions upon the uniqueness and stability of the resulting equilibria, and he proposed the startling answer: no, outside of some trivial and unavailing global restrictions. The effect this had on Debreu's older "structural" program has been nicely captured by his German protege, Werner Hildenbrand:

When I read in the seventies the publications of Sonnenschein, Mantel and Debreu on the structure of the excess demand function of an exchange economy, I was deeply consternated. Up to that time I had the naive illusion that the microeconomic foundation of the general equilibrium model, which I had admired so much, does not only allow us to prove that the model and the concept of equilibrium are logically consistent, but also allows us to show that the equilibrium is well determined. This illusion, or should I say rather this hope, was destroyed, once and for all, at least for the traditional model of exchange economies. I was tempted to repress this insight and continue to find satisfaction in proving existence of equilibrium for more general models under still weaker assumptions. However, I did not succeed in repressing the newly gained insight because I believe that a theory of economic equilibrium is incomplete if the equilibrium is not well determined. (1994, ix)

This impasse is something more substantial than the sorts of obstacles that are periodically met in the course of any vibrant science; these results have been seen as damaging precisely because they raise the question of whether the Walrasian framework is the appropriate mother-structure for the elaboration of mathematical economic theory. The Bourbakism propagated by Cowles had identified neo-Walrasianism and good economic theory, for those trained in the late 1950s and through the 1960s neo-Walrasian theory had become conflated with the very standard of mathematical rigor in economic thought. Indeed, this defined the Cowles program from its inception: why precisely should the Walrasian framework be taken as the sole "structure" from which all mathematical work should depart? And just what was the "correct" Walrasian model? Was it the one actually found in the texts of Walras, or Pareto, or Edgeworth, or Hicks, or Allais? Or to put it in Saunders Mac Lane's terms, was it not better to make a case for the "right" level of generality, than claim one had attained the maximum level?

The answer for Debreu, just as in the case of Bourbaki, was that rigor was more a matter of "structure," of style (and politics, as Mandelbrot rightly insisted) and taste; but ultimately, styles and tastes change for reasons that can only partly be accounted for by the internal criticisms generated by the activities of the closed community of mathematicians. While Debreu hoped that raised standards of mathematical economics would put economic discourse on a more stable basis, there was never any formal reason to believe it would be so.

A Conversation with Gerard Debreu

I conducted this interview with Nobel Laureate Professor Gerard Debreu over the two mornings of 4–5 May 1992.[4] The first part of the interview on

Monday, 4 May, lasted approximately two hours. The second part, on the following morning, lasted approximately one hour. Both sessions took place in Professor Debreu's office in Room 651 of Evans Hall on the University of California at Berkeley campus. Professor Debreu's office, a small, bright, corner office was very sunny on those two days; there were windows on two sides, and a regular size desk was surrounded by bookshelves on the walls of the office. Occasionally Debreu would remove books from the shelves in order to check dates, to confirm spelling of an individual's name and where they were employed, or what their positions were, or whether so and so was the research director in such and such a year, and so on. Some of the awkward transitions in the edited transcript reflect those silences in the interview associated with Professor Debreu's checking of the particular details of his account. Professor Debreu reviewed this transcript for accuracy of the quotations, and the correctness of names, places, dates, and so on, and I edited the transcript to respect his emendations.[5]

ERW: I'm interested in focusing on several matters to which you've returned on a couple of occasions in papers you have written on the mathematization of economics. You have suggested that 1944 was a confluence of, or the beginning of, several different kinds of mathematical ideas. Those interested in the mathematization theme should attend, you seemed to suggest, to the immediate postwar period. Since these are the kinds of topics that I'd like to focus on, I hope we can explore the events which led up to that period, and then talk about how that period linked to the concerns of the early 1950s.

You wrote in one of your autobiographical papers of a school mathematics teacher who prepared you in the traditional French curriculum in geometry. The mathematics preparation that you received in Calais was rather formative of your mathematical taste. I wonder if you would be able to recall a little bit more than we find in print concerning the kind of mathematical training you had in that period.

GD: I don't remember exactly when I started being a student of that mathematics teacher, Jules Dermie. It was in high school, and it would have been around my third year of high school, probably around age fourteen. It would be hard for me to remember exactly. And we cannot check because the archives of that high school must have been destroyed in the

Second World War—Calais was in a very strategic position and I imagine that the archives must have been bombed; and I myself have not kept the class curricular records.

ERW: What kind of mathematical work generally was done? It was not up through calculus, I suppose?

GD: No, there was no calculus in the high school curriculum although I studied it some on my own. Characteristic of the French curriculum in those days was that there was great emphasis on geometry, two-dimensional geometry and to some extent, three-dimensional geometry. I believe the curriculum has changed significantly at least once since that time. Maybe now, it has gone back to what I was taught? What was interesting about that form of geometry is that it called for imagination, intuition, experimentation, and I think that it gave me an excellent education, and a very good preparation for the geometric viewpoint that I have often taken up in my work since.

ERW: My understanding is that the French tradition was to separate the physics out of the mathematics curriculum. You spoke of making a move from mathematics interests to an interest in physics, with a very talented teacher who was more interested in physics than in mathematics. This was toward the end of your high school career. How was the connection between mathematics and physics made at that stage?

GD: We must have studied rational mechanics as part of the mathematics curriculum, but it is true that at the age of about seventeen (it is easy for me to remember because I was born on July 4 which is at the end of an academic year), I had a very unexciting teacher in mathematics. It was not Jules Dermie, it was somebody else. At that time I also had a dedicated, exciting teacher of physics. The kind of physics that I studied in my last year of high school in Calais was very theoretical, as I remember. There was some experimentation, but I think it was not very significant. But the theory, I loved it. It was very classical physics. I am surprised in retrospect that while I studied physics up to the university graduate level in 1941–42, that there was not a word about atomic physics, quantum physics, nuclear physics, relativity theory; those things were not in the curriculum.

ERW: Was there any discussion of probabilistic approaches, statistical approaches, in physics, prior to your college years?

GD: I actually learned probability theory and statistics significantly later, on my own. It was not part of the school curriculum that I took. I was not required to take it and I did not at that time take it.

ERW: What other kinds of preparation would have been done outside the high school? You were preparing to take examinations to enter one of the Grand Écoles. Are there are sets of examinations that one takes in order to qualify for them?

GD: Let me explain about the French system. After the Baccalaureat, when you are eighteen years of age, you go to some special class, which in my case was called Mathématique Spéciale Préparatoire, and the following year, Mathématique Spéciale. All students in those classes take the first year and then they take the second year. At the end of the second year they take the comprehensive examinations for entrance into the Grand Écoles; if you fail at least to enter one of the schools that you want to go (there is such a wide array) you can repeat the exams and you can, even though it is rare, repeat again, but that is a sign of failure. If you fail three times it is not a good omen.

ERW: How much of the mathematics training that you have to draw on came from those two years of preparation done in physically different places because of wartime conditions? You moved around. Each of those years was done in a different place with different instructors.

GD: For the first year it was in Ambert, a very small town.

ERW: What kind of preparation did you receive? How many other boys were there? I assume it was all boys.

GD: Yes, all boys. How many classes? I believe there were three classes. One was Mathématiques Spéciale Préparatoire in which I was, then there was a year of Mathématiques Spéciale (the curriculum you take in the second year, but which you can repeat). I think there was a third class there as well, one of preparation to St. Cyr, which is the school which trains officers for the French army, and which also probably trained officers for the infantry, the armor, etc.

ERW: And the mathematics training that was provided in those classes you took there, I suppose was beyond the kind of geometric analysis you had in Calais. Would it have included analysis using say Goursat, using Picard, or was it more basic stuff, at a lower level?

GD: No, it was below that level. I do remember the text book that was used—Commisaire and Cagnac—it was a popular textbook at that level at that time. In any case, the course I took was taught by a Professor Croissard who had been called back to active teaching service because of the war. Since so many of the younger people had gone into the Army, the call was made to retired persons like him. Naturally he was very competent. He was also full of enthusiasm, which is important to have as a teacher. That is what happened to me in high school: Dermie was very enthusiastic about his subject, sometimes too enthusiastic for a student who was not doing what he was supposed to do. He could become angry and almost violent. Though he did not hit students, I do remember one episode when he literally threw a student out of the class. And I do not know what the student had done!

And as you said earlier it was a boys' school; however, in the last year of high school students divide basically into two branches: philosophy and mathematics. And the mathematics course was not offered in the girls school in Calais and so all of them took the same classes in mathematics with us boys. But in Mathématiques Spéciale and Mathématiques Spéciale Préparatoire it was only boys, as I remember.

Now the school in Ambert was not intended ever to have that kind of special class. It was a high school regularly, in a very small town in the Massif Centrale. I was supposed to have taken the first Mathématiques Spéciale Préparatoire in Paris, but because of the war conditions, I had to go elsewhere—Paris was thought to be a prime target, so I ended up in Ambert. I think that that class folded up after I left, so it was just a one-year class there. I then had to do the second year. Grenoble had a Lycée, and it had every year, normally, both the Mathématiques Spéciale Préparatoire and Mathématiques Spéciale. It may have had more students then usual at that time because a number of people had gone south because of the war, like myself. Grenoble was in Free France at the time. So when I was in Ambert, France was divided by the occupation forces and I found myself south of the line and so legally I could go only to some Lycée also in the south, Grenoble in that case. I had no reason to regret my choice, because at the end of my year in Grenoble I was admitted to the École Normale Supérieure in Paris. At that point then I was faced with either the red tape of obtaining a visa, or crossing illegally to get to Paris, which is what I did.

ERW: Had your family followed you south or did they remain in the north?

GD: No. I was a boarder in Ambert. Almost everybody was a boarder. There was no hotel, and in any case students . . . I would have been a boarder in Paris had I gone to Paris, but I was a boarder in Ambert. And the École Normale Supérieure provided that I was a boarder too.

ERW: Were any of the instructors in the special classes linked to the emerging Bourbaki group?

GD: Not at that stage. Later, I met some but I was then in Paris.

ERW: So there would have been very limited toplogy, set theory, and abstract algebra at all in the preparation. What kind of examinations were there for the Écoles? Do you have any memory? I ask because you really would have spent about two years thinking about these examinations, and the kind of questions and the kind of way of thinking that those examinations would coerce would likely have an effect on your subsequent work.

GD: Your question puzzles me. I know the examinations were long, maybe six hours. There must have been four different examinations; but what the subjects were, I don't really recall. They were very classical mathematics, I'm sure, because there was no study in those classes (Mathématiques Spéciale Préparatoire and Mathématiques Spéciale) of topology, set theory, or abstract algebra. None whatsoever. It was a study of the calculus, infinitesmals, a lot of geometry, three-dimensional geometry—those were the main subjects that were studied. And so the examinations must have been centered on those subjects, but what they were exactly I am not sure.

ERW: Would applications have been drawn? You have mentioned the possible preparation for St. Cyr; in that class there would have been, I assume, engineering applications and examples.

GD: Let me say that I knew nothing about what went on in the St. Cyr class, people who were intending to become professional army officers, they were completely different, so I don't know what they did. In general, concerning applications of mathematics, I recall that there was one of the exams for entrance to the École Normale Supérieure, it must have been in physics, but it was again purely theoretical physics. There was no chance in that, if it was a six-hour exam, no chance for experiments.

ERW: Do you remember any of the texts that you would have used or any of the things that you might have read at that time in preparation?

GD: In the course that I took, the popular textbook in those days, and I'm sure it was altered completely since, was by Commissaire and Cagnac. Those were very good books; I read them for pleasure, not just out of duty but for pleasure.

ERW: I'm also interested in trying to understand and interpret mathematical taste, in particular, your mathematical taste and interests. There are, besides mathematics, the other kinds of things that go to form a mathematical taste; they come in from other kinds of directions whether they be aesthetic, or philosophical, or scientific in different kinds of ways. I get a sense that the formation of your mathematical taste was not so much out of an experimental scientific traditions as much as it was, even before the École Normale Supérieure, was purely mathematics together with other things perhaps. Were there aesthetic tastes that were joined in your view? You've often written about mathematics in language that would suggest that the Bourbaki understanding of what mathematics is all about struck some very resonant chords with you by the time you met Cartan, and I wonder what the fertile ground was aside from just the connection. Had you studied music, had you studied art, formal or informal?

GD: I am not an artist. I may have tried, but I found out that I was not gifted, so I didn't pursue it.

You must understand that there was a fundamental difference between the two kinds of mathematicians I met: before 1941, those were teachers not research workers, and after 1941 when I was at the École Normale Supérieure, they were first of all research workers and secondarily they were teachers.

ERW: You came to Paris from Grenoble. You obviously did well on the examinations. You wrote somewhere that there were only twenty mathematics students at the École Normale Supérieure and thirty in the humanities.

GD: The sciences were divided basically between mathematics on one hand and physics and chemistry on the other (the two went together) and there was a third possibility (but very few students went that way), that was biology. And I imagine that in our group maybe only one or two went the way of biology whereas the division between mathematics and physics and chemistry was about even.

ERW: Was that division determined upon entrance or did that emerge over the period?

GD: All students take the same examination when they enter the school, and then we decided which way to go.

ERW: Was it a three-year course?

GD: It depended. In mathematics it was normally three years and in physics I think it was four. And at one point I thought I wanted to take my distance from mathematics because it was very abstract, and as I wrote somewhere else I was interested in several directions. One of them was economics, as you well know, but one was astrophysics, though I did not go very far. The problem in astrophysics was that first of all, the faculty at the University of Paris was depleted during the Second World War. I think some of them were Jewish and it was unwise for them to stay in Paris. And others were communists (and some were both) and that was certainly the case with Joulliot. He was I don't know where but he was certainly in serious danger. What he did during the war I have not checked, but I don't think he was around. So what happened in astrophysics is that when I looked around, I found—maybe my search was not long enough, deep enough—but I had the impression that there was no faculty so it was not a very promising field because I would have had to study entirely on my own to stay in that field. So it would have been difficult. So I did not stay very long with that.

ERW: When did you first have contact with the Bourbaki view of mathematics? Did you meet Cartan immediately at the École Normale Supérieure?

GD: I am not sure. It was very likely so I would say I met him in 1941 but I wonder I may be off by one year. In any case I was aware of what Bourbaki volumes had already appeared, which in fact by 1941 was very little, I think only two volumes. And even then, one was a summary.

ERW: And that summary was very different from what did emerge. There is in the mathematics a tension between what some people have now taken to describing as *structure* and "structure" as two different notions of structure. I'd like to perhaps to return to that after a little bit.

The mathematics instruction probably would have been, or I assume it was, a core instruction for all of the math and science students probably on entrance and then some kind of bifurcation perhaps later. The initial instruction in mathematics would have been perhaps in a lecture form, or was this done by Cartan?

GD: It's very strange. Again, it is unique. If you take another Grand École like the École Polytechnique, they have all their teaching within the school only for students there. Not at the École Normale Supérieure. It is close to the Sorbonne, geographically close, and we were supposed to take the standard courses at the Sorbonne. And what we had at the École Normale Supérieure was very small seminars; that is where we were taught by Cartan. There was no fixed curriculum, and it was attended by about ten people, whereas in the fundamental courses at the Sorbonne the attendance was at first in the hundreds. I do remember a course taught by the physicist Yves Rocard, I believe he is the father of a prime minister, and I found that since there were so many students (and the lectures were available in writing) that I stopped going to them altogether. What was lacking in them then was the enthusiasm that Cartan generated.

ERW: In a small seminar taught by someone as competent and organized as Cartan, did he use the Bourbaki material that was being produced over that time?

GD: No, it was very erratic, he talked about various things. It was not related to the two volumes that Bourbaki had produced, not at all.

ERW: Do you remember any of the other students? Did they become mathematicians? Were they as taken with the material as obviously you were?

GD: Some of them became university professors of mathematics and one of them became a probabilist, one a biologist, he was the only one who became a very productive biologist. My recollection of those days are very imperfect. The objective at the École Normale Supérieure was basically to produce teachers of mathematics; and that was understood in the days when I was there, to mean teachers of mathematics at the Mathématiques Spéciale Préparatoire and Mathématiques Spéciale level. Students had to make decisions then whether they wanted to become teachers or research workers, and some of them went one way and some went the other. I do not know whether the decision was made as we entered, or whether we discovered two years later we might want to do research.

ERW: At that time were you thinking of becoming a mathematics teacher?

GD: A university mathematician (teacher not research worker), yes, that was my intention when I entered. It seemed to be the natural development.

ERW: Though your family seemed not to have been, at least you have not mentioned—academics.

GD: No. But it seemed to be a very natural development given the past five, six years. After a year or so (I entered in the fall of 1941), I began to wonder whether mathematics being at that time, becoming very abstract, and the influence of Bourbaki was not so very dominant as it later became (though maybe I anticipated that development), and I had to decide whether I wanted to spend my entire life doing research in a very abstract subject. You must also remember that during that last year of high school when I was influenced by my physics teacher I had thought that physics was going to be my field.

ERW: The period in Paris, then, marked a transition; though you studied mathematics intensely and almost exclusively over that period, you were not pleased with the thought of becoming a professional mathematician.

GD: Yes.

ERW: Do you remember other things that you read before the copy of the Allais book came to you.

GD: When I became interested in economics as a possibility (as before I had become interested in astrophysics) I got hold of the standard text studied by students of economics at the university. I don't remember who the author was but it was very nontheoretical (somebody I never met). I know the textbook was popular then—I don't believe I have kept a copy— but in any case, my first impression of economics was very disappointing because I was coming from a world of very sophisticated and rarified mathematics and found only a very pedestrian approach to economics.

ERW: I want to return to the question of the kind of mathematics that you would have been studying in the seminar. Had you read van der Waerden's algebra by that time?

GD: No. I knew of its existence, I may have read a few pages, but I had not read the book.

ERW: You have said that the Cartan seminar was a very different intellectual experience from the rest of your mathematical education. What did this consist of? Would it have been lectures, discussion of problems, reading of classic papers?

GD: At the university it was lectures, fairly polished lectures by . . . and I do remember the names of some of the lecturers. One was, as I said, Rocard whose lectures I did not attend eventually except for the first two or three. But I faithfully attended the lectures by Garnier, and Garnier taught differential geometry. Valiron taught classical analysis, and later on I took lectures by Gaston Julia on Hilbert Space. I'm sure I have taken others. And then in the seminar it was a mixed bag; we occasionally had a lecture by Elie Cartan, the father of Henri, who was of course already at that time a very revered mathematician. We had a lecture by De Broglie, the physicist, Nobel Prize winner. So the seminar was a little of different things by different people. Henri Cartan was still young, and did great things later, and the seminar was simply supposed to review, and it did that; it was also to give us a taste of a variety of mathematical researches, and no text was used. On my own, I read most of Goursat.

But remember that I was in a very bizarre situation because it had become clear to me that I would not be a professional mathematician. It was late when I decided that, and my career was disturbed by events of the war. I was supposed to take the final examination in the spring of 1944 after three years, but that was exactly when D-Day occurred. I took that examination eventually in 1946. So I finished my studies two years late: I was supposed to finish in 1944 and I actually finished in 1946. So it is difficult to say what would have happened if D-Day had not occurred because that two-year delay (part of which time I spent in the French Army) gave me a chance to get much better acquainted with economics than I would have been if my curriculum had followed its normal course.

But in any case when I took the aggrégation, it was a pure mathematics examination. It was a somewhat bizarre situation. It was an examination of a somewhat scholastic nature, which was all the more so for me; it was very classical, whereas Cartan in particular had done contemporary mathematics, which was not the case with the aggrégation, so I had a number of problems.

ERW: The move to economics begins looking more unusual. With Tinbergen the commitment to socialism as a young man was part of his move in the connection from physics to economics. In your case, the move to economics seems to have been a move away from pure mathematics: a second choice perhaps to doing astrophysics but that was a field in which you found no one to study with. What were the kinds of connections with the choice looking back on it? Was it that your family was in business, or was it just an interest in these ideas?

GD: It was the times and the events. And also to a large extent it was pure chance because Allais had sent his book to a friend of mine, who was a humanist, who was the president of his class. He was actually not in my own class but one year after, but we were friends and he gave me his copy. I suppose otherwise if I had persevered in my interest in economics that I might not have been aware of the Allais book for months, and maybe by then it would have been too late.

But one part of my interest in economics—although it was not too elegant a field—was simply that the war economy in France was special. We believed, though we found it a long time coming, that Germany was going to be defeated. It was clear that there would be a lot of reconstruction in particular in France. There would be a lot of reconstruction work to do after the war and it proved to be the case. And that may be why I came to know people like Pierre Massé, who was at one time president of Electricité de France, but who also wrote a book on stock management. I came to know him very well and I saw him regularly until his death. He was succeeded as president of Electricité de France by a friend of mine, Marcel Boiteaux, who also had his career disturbed by the war. He was an officer in Italy somewhere and later on after the aggrégation cast his lot with Electricité de France. We shared the extraordinary story of the coin-tossing for the Rockefeller Fellowship. So a number of random events were surely very important.

ERW: The book by Allais arriving in your hands at a very crucial kind of moment or at least one that was, say random, unpredictable kinds of branching of choices. My understanding is that the book was more or less just sent around. It was privately printed. It was very much outside the official channels. In retrospect, it's remarkable that it was able to be printed under those wartime conditions. That it was not only printed, but that it received any hearing at all given the number of unusual books that drop out of sight, leads me to ask how was this book taken up? It was handed to you with not much understanding of the connection to what would become your own interest. The mathematics was not up to the kind of standard of course that you were able to deal with. It was very primitive mathematics from a Bourbaki point of view though it was more sophisticated than Hicks certainly. But was it an approach that you found congenial? Did it touch your desire to frame the economic questions in a rigorous way?

GD: It was several things. First of all I saw that mathematics could be used in economics in a rigorous way, even though it was not the kind of mathematics I was most fond of. And maybe I felt that there was a lot to be done, with more sophisticated mathematics in economics. And as I was saying earlier, my interest in economics wasn't ready-made. I became interested in economics in 1943 or maybe a year before. And the circumstances were such that Bompaire gave me that book let us say around April of 1944. But I think that the circumstances were not sort of ideal for me to read it just then—recall that D-Day was 6 June. It was only in September, I suppose, when it was clear first of all that I would not start another academic year normally. Things were too chaotic in France and it was then I think that I became serious about economics; I may have looked at it when Bompaire gave it to me but I did not study it. And you are right, it was that probably few people got hold of copies and my guess would be that most of them must have been repulsed by all the characteristics of the book; it was lengthy, technical, etc.

ERW: And the economists in France, especially those who were outside the kinds of Polytechnique tradition of doing economics without being economists, would have looked at the book as something that came from Mars.

GD: Absolutely.

ERW: How did Allais take this reception?

GD: You must know that students of the École Polytechnique, and Allais was one of them, are looked at with suspicion by people such as professors of economics who have a very legalistic education. And for somebody who does not know mathematics it is impossible to know whether that book was serious or the product of one more mathematically mad person. It is very difficult to know. So I would imagine that the book was poorly received by the official economists. Have you met Allais? He is not an easy person. He has a gift for antagonizing people. And in fact his great disappointment, I know privately, was that he was never appointed professor of economics. So he taught, I think, during most of his career if not all of his career, at the École de Mines, where he had a regular post; I have had as Ph.D. students here those who have taken his course at the École National Superieur de Mines.

I must have met him for the first time after the aggrégation, in 1946, because when I came back to Paris after my experience in the French Army, the aggrégation became my main concern. I left to go to that examination.

ERW: Did you meet him with the idea of perhaps studying with him, or making connections to a way of thinking about doing economics for yourself?

GD: I think it was clear at that time that I was ready to cast my lot in economics and I certainly believed at that time that he was the most creative economist in France. Very few economists used mathematics in economics. I do remember François Divisia was one, Rene Roy was another and I think they were the only representatives of the school which I was interested in. Both Divisia and Roy were significantly older than Allais so Allais had the advantage of youth, enthusiasm, and energy.

ERW: You said in a couple of places that after you had read Allais, or that as you were reading him, Divisia's book was the second book in economics really that you had read.

GD: Right. I believe that Divisia was Allais's teacher at the École Polytechnique when Allais was a student. (Looking at Divisia's book.) It is in bad condition, not entirely because I read it with care but because it is paperbound and because it is old. But I read it carefully.

ERW: That was the second economics book you read. Looking back do you have a sense that, from the perspective of economics, you were an outsider in very fundamental ways. That this group really consisted of Allais and Divisia and Roy and then you and Boiteaux, but it was a small group. In looking back to what it must have been at the time, it was a very large intellectual risk that you took to cast your lot with something that was not only not popular, but was outside all the established ways of thinking. Did you have a sense at that time that this was an intellectual risk that you were taking?

GD: Probably. But I was very young and had no responsibilities. I was not married. I had lived through a period, the war period where the risk was high, yes. I was sometimes in danger of no less, sometimes, of my life, so that seemed to be a small risk. The times made me more risk-loving than I later became. And it was, I think, clear to me that it would be difficult if not impossible to get a chair in economics at the University of Paris. And I was right about that, for theorists were not appointed as professors of economics at the university until very recently. That much seemed clear. Indeed Boiteaux was in a somewhat similar situation of having a totally unexpected career; he cast his lot with Electricité de France—a very large organization—of which he eventually became the president. So my conception of what would become of me was very vague. It was more or less clear that I would be an economist, and that I would not be a professor of economics at a university. Possibly I had in mind vaguely jobs like, let us say, Boiteaux's job. After all, I supposed that a mathematical economist would have been also in other industries in France outside the university. But this was, I am sure, very vague in my mind. There were organizations, European organizations, maybe organizations associated with the United Nations, possibly the World Bank. I do not know exactly. Actually this career was not designed, but it was clear, as I said, that I intended to become an economist, but extremely unlikely, I would say impossible, to become a professor of economics in a university.

ERW: I believe it was after the aggrégation that you had what you describe as a wonderful period at the Centre National: that you were able, for almost a two-year period, to explore the kinds of issues that were beginning to engage your interests. In some ways you seem to have laid out a research program in that two-year period. Many of the kinds of themes that engaged you over this next fairly longer period of time began to take root at that stage. Is my reading of that fairly accurate?

GD: It is true that I had extraordinary freedom for approximately two years to two and a half years at the Centre National de Researche Scientifique. On a research program, I am not sure. Certainly I had a lot of reading to do because I was basically learning the subject. And I learned it, I am sure, in a very unorthodox unsystematic way, but I read at that time a number of the classics. I remember Frisch, Pareto, Walras, Gide, and Rist, and many, many others. I have never attempted to make a list of this.

ERW: Were you reading the *Econometrica* back issues?

GD: No, probably not.

ERW: Do you remember having read Evans's book on mathematics in economics?

GD: Yes. I read it in Paris before I went to Salzburg. One of my mentors/sponsors in Paris was Georges D'Armoire, the probability theorist, and he was by no means an economist but I think he happened to have a copy of the book at home so he gave it to me to read. I was not very enthusiastic about the Evans book; I met him later but his work in economics did not seem to me to be very important. I read a strange collection of books. I read Zeuthen's book in those days; it was a thin book but surely that was a strange way to be introduced to economics. Zeuthen's book was mentioned later in Harsanyi's paper, when he compared the two approaches to the Nash solution and pointed out that Zeuthen's approach formally led to the same solution as Nash.

ERW: Were you reading any of Samuelson's work at that time?

GD: I suppose I became aware of Samuelson in Salzburg in the summer of 1948 at that seminar organized by the university where Leontief and Bob Solow came over; Solow was a little young but was very aware of the work of Samuelson. I believe it was also at the Salzburg seminar when I first saw the book by von Neumann and Morgenstern, at least that was when I started reading it. France was cut off from the rest of the world in the war, of course. But after the war it took some time to reconnect, and the library system in any case in Paris is not very efficient, whereas in Salzburg they had made an effort to import, to make available, a number of these most important, most recent books in particular.

So the Salzburg seminar was a very important time, not only because it put me in touch with some American economists, and some of the literature of the late 1930s and early 1940s, but also it gave me a chance to see young intellectuals from other countries: Denmark, Norway, there was even, I think, at least one German.

I believe that I saw the von Neumann-Morgenstern book for the first time in Salzburg. I read it essentially at the beginning of 1949 when I was a Rockefeller fellow; I spent the first six months of 1949 at Harvard and I remember that while there I really studied carefully, not the entire book by any means—it is an undigestible book—but at least a very significant part, the main concepts.

ERW: At that time what did you feel was the contribution of the book? Do you recall a kind of first reading of it? Was it the incendiary introduction or specific kinds of problems that were taken up?

GD: It is difficult to recall. It was a new subject, and it attracted attention. I liked it because it was in accord with my mathematical taste and I was fairly deeply impressed by the formalization of games, the concept of information, the minimax theorem. I was indifferent about their central concept of a solution. It proved to be a very useful, usable concept.

ERW: But the book was not really discussed at Salzberg?

GD: No. It was a very diverse group there; there were only maybe half a dozen economists. There were other people interested in constitutional law, literature, history, etc. In that group of young French economists, you mentioned yesterday, Boiteaux and myself, I do remember at least two others who came fairly regularly to the meetings with Allais and Roy, and they were Pierre Mayer (I am not sure about the first name). And the other one has a similar name, Maillet, and there must have been others as well. But I remember especially those two. Maybe one person who came less often was Charreton and I saw Charreton again at the World Congress of the Econometric Society in Aix-en-Provence, and I think he had become, in the meantime, an economist for one of the French petroleum groups. I am not sure what the others did. I think that Mayer was at the Ministry of the Treasury, and I suppose that Maillet worked for the European Community on steel and coal, but I am not even sure. So they have not followed the scholarly route, and Charreton has not either.

ERW: Was it this connection then, in the summer of 1948, that led you to pursue the Rockefeller Fellowship?

GD: It was independent of that. It may have reinforced it, but I think it would have happened in any case.

ERW: The sense that there was perhaps a community of like-minded individuals then was beginning to get reinforced through your reading though you didn't see very many of them in Paris, because they were there in other places. Did you have a sense that your own background was dramatically different from theirs? You mentioned Leontief and Solow were more traditionally trained.

GD: Yes, but not dramatically different from me. They were certainly much closer to me than the typical French economist. Certainly in France I had the impression, which was accurate, that I was in a very small minority and that would not be the case in the United States, from what I gathered. But this does not mean that mathematical economics was well accepted in 1950 when I began.

ERW: You began that career in the United States after some traveling, and a short visit to the Cowles Commission. All of the kinds of memoirs and stories that have been done on Cowles still don't seem to capture the richness of the experience and the nature of the connections. Your own look back at Cowles's mathematical economics conveys almost a sense of wonder at the sheer number of individuals who passed through Cowles, who made the kinds of contributions that we now regard as having been so very important. It is really a remarkable configuration in that Chicago period. I wonder if I could get you to try to go back to that period to when you first got there, to seeing some of the individuals. Who would have been those who introduced you to the others and made you feel welcome?

GD: I saw the Cowles Commission first when I visited in the summer for a couple of weeks. That must have been in the fall of 1949. I had the Rockefeller Fellowship from the last days of 1948 to basically the time when I joined the Cowles Commission for good on 1 June 1950. So on that fellowship, during that year and a half, I had visited different places, Harvard for six months, Berkeley for a few weeks, and then Chicago. And indeed in Chicago I found an atmosphere that was the most remarkable of the several universities I had experienced; Harvard was really very encouraging, infinitely better it seemed to me from my viewpoint than Paris, but the Cowles Commission was even better. They asked me to speak. Who were the most obvious people at the time? Koopmans, the director of research, Jacob Marschak, and among the others I think that Leo Hurwicz was there when I spoke, I don't think he was a member, I do not know. Maybe he was back in Urbana so it was not a major trip for him to come to Chicago.

ERW: Koopmans was the research director then?

GD: I think so, yes.

ERW: And Koopmans also was an individual who came not from economics but came from physics and had done war-related work in more mathematical economics. He was an outsider to economics in those ways. Marschak was more the economist but he was also an outsider. Cowles seemed to collect a group of individuals who were not wedded to seeing economics in the same way that the presidents of the American Economic Association at that time perhaps would have seen economics.

GD: Absolutely.

ERW: Was there a sense of a small group banded together trying to create something different or was it more, "We are doing what we are doing"?

GD: To me it seemed to be fairly unique in the world in those days. There were mathematical economists elsewhere, there was Morgenstern in Princeton, and in Cambridge, Massachusetts there were people like Samuelson and Leontief. Solow was a young theorist, and some of the mathematical economists I knew spent a significant part of the summer at RAND. I did not do that and that may be due to some extent, but not entirely, because I was not a U.S. citizen, and RAND was doing a number of things for the army.

ERW: Was the work at Cowles directed specifically at a certain problem?

GD: When I joined the group in June 1950, it seemed to me to be a very theoretical group. In particular, the Cowles Commission monograph on estimation had by then been written and published. But Koopmans himself made a fairly drastic change because in the days when this book was developed he was deeply involved in econometrics. But from the time when I knew him, he was never, I believe, working actively on estimation methods, and he had become an economic theorist. That can easily be checked in his bibliography, but I believe that had happened already by the fall of 1949, and certainly when I knew him well after June 1950, he was doing work exclusively in theory. He worked on some of the applications of theory, but that came much later. And Marschak I would say was basically the same. In those days the Cowles Commission monograph on activity analysis was not yet published though it was published I believe shortly after I arrived.

ERW: That was the result of a conference that was held. You've referred to that book as probably the most important book in defining the mathematical economics areas. I've come back to that book time and time again because sequences of citations and references seem to flow through that book perhaps more than any other that was written at that time. The conference I believe was in the summer of 1949. The book shows that many of the references and authors there went back to von Neumann and Morgenstern and also to von Neumann's 1936 paper to cite precedent for approach by convexity to optimization issues. It really was the Koopmans volume that focused these ideas. Was it seen at that time at Cowles as a significant a change in point of view? Was it a revolutionary program that one was engaged in or was it simply, "This is a new language and we're just talking it"?

GD: It is hard to say because when I arrived it was already somewhat in the past so I could judge the excitement only a posteriori. But I think that it was considered a very important event in history, the intellectual history of economics. Publication of the volume was thought to be important. And in that book in particular there was the paper by George Dantzig, the first publication of the simplex algorithm, and in general, mathematical rigor was throughout accepted as a premise, by which I mean full mathematical rigor, which was not so common in economics then. There were papers in that book by Gerstenhaber, Gale, Kuhn, and Dantzig, of course, and by a number of economists like Koopmans and Marschak who were to some extent outsiders even though they were professors of economics in Chicago. I am sure there were frictions with Milton Friedman for many reasons not only concerning the use of mathematics, the general abstract approach, but the ideology also.

ERW: There was the issue of general equilibrium analysis itself which came out in Friedman's review of Lange earlier, though Lange was Chicago at that time. I want to come back to a little bit earlier period, to the 1944 von Neumann-Morgenstern book. There was a sense in which that volume had no effect at all on economics until almost the 1960s. You point to the Nash paper as making the difference in the reception. Others have suggested that the book lived on in an underworld almost, until the theory of cooperative games was rescued by Shubik in his Edgeworth paper. I've also been engaged by an argument which suggested that the book was met with suspicion by economists because both of the individual promulgators of the theory, von Neumann and Morgenstern, were alienated from economics and were in many ways quite hostile to what was considered to be acceptable economics. And for that reason, neither of the authors was willing to engage the economics profession in taking up the book and that it had to be left to others. Von Neumann saw the book as a revolutionary tract, and Morgenstern, as an Austrian economist, was engaged in the project because it could be critical, provide a critical perspective, on standard economics. The book was almost too frightening at some deep level for professional economists to engage and it could only come into economics through the interests of individuals who were not trained as economists. And it took a while then, so it would have been engaged first by Cowles in some ways. It is a peculiar line of argumentation because we've now domesticated that book and see it as part of our tradition. But at the time, did it seem continuous with the kinds of things that were being done at Cowles, or was it a collection of tools that could be taken up to solve particular problems?

GD: My feeling is that it was accepted fully at Cowles. There were two favorable book reviews, one by Hurwicz, the other by Marschak. There was also disappointment at first because for many years, not much came out of game theory as you noticed and the mathematicians who worked on the theory of games extended the minimax theorem with more and more general conditions but that was certainly not essential and for a small number of theorists, the paper by Nash was indeed important. The 1930s papers by von Neumann and Wald were well known at Cowles but that small group at Cowles was not representative of the profession. It was exceptional in many ways. The group lived in a universe of its own which made it possible for them to live very comfortably with ideas that were not orthodox. There were many fewer colloquia in those days than there are now, so there were fewer opportunities to be confronted with other viewpoints and when the colloquium was on linear programming there was no reason to disagree. Incidentally, von Neumann played an important role in the development of linear programming. You will find in his collected papers the paper he wrote on the duality theorem which was never published.

ERW: Von Neumann was increasingly taking up a very constructivist mathematics program in those day where game theory strategies connected to his ideas on automata theory. The theory of games became lodged at RAND in some ways because it was taken up by the military applications at that time. Was there much traffic between Cowles and RAND? You said that you yourself were not able to work there because you were not a citizen.

GD: I do not know whether this was the reason but I remembered then thinking that it was so. Perhaps I suspect it now more than I did then, but I may be wrong. I do not know who from Cowles went to RAND in the summer.

ERW: Let me ask about some work that you didn't do. The questions of dynamics were of concern to Samuelson in the late 1930s and early 1940s; his own intellectual history reflected a very different kind of mathematical career and training. The issues of dynamics didn't seem to have engaged you, although you had a knowledge of differential topology sufficient, even at that time, to have perhaps done things very differently. Is there any particular reason why those questions of dynamics, questions certainly alive at Cowles at that time, or even a little bit earlier, did not engage your energies?

GD: The interest at Cowles was probably earlier than my time there. When I was there, from 1950 to 1960, those questions seem to have been totally absent, unless I am mistaken, but I can't recall an instance.

ERW: There were the Mosak and Lange books of the early 1940s.

GD: But it was in the 1950s when I had the fellowship at Cowles, from 1 June 1950 to the late 1950s. In any case I had my own reservations about dynamics in spite of the fact that I had studied classical mechanics, studied it quite extensively in fact; therefore, it would have been I suppose possible to transpose the ideas but I thought that the whole question was very facile, and that in economics one did not specify, then test, the dynamic equations that were so easily taken up because of the analogy to classical mechanics. So I was very, always very, suspicious of dynamics and that is a view that I have held very consistently.

ERW: So these parts of Allais's book concerning dynamics, and then when later at CNRS, when you would have read Hicks and so on, these ideas didn't engage you intellectually?

GD: It was not that at all. I thought about those questions of course as every economist must, but it seemed to me that the contributions made were not important. For one thing, when you are out of equilibrium, in economics you cannot assume that every commodity has a unique price because that is already an equilibrium determination. The process should in particular take into account the fact that the same commodity has different prices at the same time so that makes dynamics among other things very hard. And you have to recognize that in classical mechanics you have a simple correlation: force is proportional to acceleration. We have nothing similar in economics. So I have always been very distrustful of dynamics and I have mentioned it rarely. I don't believe I very specifically even mentioned it when I showed that any excess demand function can be obtained from a suitably chosen economic system, and that means that if you take a dynamic equation of a classical form, dp/dt as a function of p, that means that you may have a vector field on the unit sphere which is completely arbitrary, therefore you may have any dynamic behavior.

ERW: What changed when Cowles left Chicago?

GD: For me, it was different before 1955 in Chicago and after 1955 in

New Haven. Certainly the group of people were different. But the atmosphere was different. It was a different director of research before 1955. After 1955 (I am not sure about the dates), Tobin was director of research until I left in 1960, spending 1960–61 on leave at the Stanford Center for Advanced Study.

I joined Cowles, as you remember, on 1 June 1950. I had visited the fall before for a few weeks. It was the time when I presented a paper there, and then they offered me a one-year visiting appointment which was transformed into longer and longer appointments. Eventually I was an associate with them for eleven years.

ERW: In the Chicago period, were you doing any teaching?

GD: I was doing some teaching. I was not a member of the department. That was one thing that happened at the time of the move. In Chicago the Cowles Commission was not administratively part of the university although it was housed in a university building. Whereas in New Haven, the Cowles Foundation became a part of the department. In Chicago, a number of people had appointments as well in the department: that was the case with Koopmans, Marschak, and a few others. But more junior people like me did not have a departmental appointment. Nevertheless, I did teach a graduate course in mathematical economics; in Chicago they were on the quarter system, and I must have taught for one quarter a year, and not every year, but I really taught what I was doing research about. It was very stimulating to have to systematize my ideas to interest students and certainly it was an important part of the Commission.

ERW: Were there issues about the level of mathematical sophistication that students were presenting? There seemed to be a change in roughly that period, in the early 1950s, regarding the mathematical preparation of economics students. It was an open question and was contested within the economics profession, with the call of the AEA Commission suggesting that graduate students should become better prepared. Was this a lively topic of discussion?

GD: Not for me, but I did not attend department meetings, since I was not a member of the department. The teaching of economics was not one of the most important issues at Cowles; certainly I did not experience that problem of preparation in my own teaching because the students had to be very well prepared because I taught them a difficult course.

ERW: I am interested in the level of mathematical sophistication of your colleagues at Cowles, particularly Koopmans and Marschak and so on. Could you talk a little bit about their own mathematical perspectives, and the degree to which you felt you were able to be speaking the same language. Koopmans and Marschak were both trained of course in very different traditions.

GD: I do remember that Koopmans was definitely better trained in mathematics than Marschak. But Marschak was amazing because he went to every seminar and understood if not every detail, all the main ideas. He was always interested. Koopmans was very much at ease with all the mathematics that were used at the time—they were less sophisticated than nowadays—but I do remember that he was not familiar with the definition of a Banach space, because somebody had used the concept of a Banach space, and he asked for a definition, so I imagine that he was not familiar with infinite dimensional spaces; he did not use the idea in any of his papers. But as you can tell from his writings, he was well prepared mathematically and he put emphasis on complete rigor.

We had also as consultants some young graduate students or assistant professors, young mathematicians. Morton Slater was there for a year after I joined. Herstein was there too. I wrote a paper with him on non-negative square matrices. He joined the university in 1948, joined the staff of Cowles Foundation in late 1951, as a research associate and assistant professor and at the same time was appointed a research associate in the department of mathematics.

ERW: Was John Milnor there in the same kind of position?

GD: John Milnor was never formally associated in any way. He was, I believe, at Princeton at the time. He came to Chicago for very short visits, according to my recollection, in the early 1950s. He wrote that paper with Herstein, where they started collaborating, it may have been at RAND. In the Herstein and Milnor paper it says the report was supported by the RAND Corporation. It does not say when they started the collaboration. I shared an office with Morton Slater, and possibly, yes, two other people in Chicago. McKenzie was a visitor, I think, in 1950–51; that is my recollection.

ERW: Kenneth Arrow had already left for the west coast at that point.

GD: Yes. We met for the first time at Stanford at the end of 1952. I do remember very well that I took the train, it was how we traveled in those days. It was from Chicago to Palo Alto, it may very well be, because it went around the south of the bay. It was shortly before Christmas and I presented a paper at the meeting of the American Economic Association between Christmas and the New Year, I believe, in Chicago.

ERW: Arrow's own mathematics training had been primarily undergraduate, and then he moved directly into a program in economics, though his interest in mathematical statistics had led him to Hotelling. So his mathematics was not nearly as sophisticated at that time as your own.

GD: In a way that is true because I was trained after all as a mathematician even though it was against my own wishes. Arrow wrote his dissertation with Hotelling at Columbia and I imagine the emphasis was on statistical methods; at that time, I suppose that Abraham Wald was also at Columbia. So here was an eminent statistician at that time, who argued the question of equilibrium in economics. But Arrow probably did not know about it because it was a short phase in Wald's career, because, as you have read, his history was as a refugee from Romania.

ERW: Wald had been briefly associated with Cowles in the Colorado Springs period before he left. Did he have any connection on a continuing basis?

GD: No, I don't think so. I don't believe I ever spoke to him. I heard him lecture once but it was on mathematical statistics and I believe his association with the Cowles group was very short. I merely mention him in my paper "Mathematical Economics at Cowles." I wrote there that "Abraham Wald who was appointed as a research fellow for one year in July 1938, actually left for Columbia University in September."

ERW: Can you talk a bit about Marschak and Marschak's own mathematical tastes, his problem taste, and his involvement in the direction of theoretical research at Cowles. Except for the discussions in Cowles retrospectives, Marschak is a greatly underreported individual. He is one of these figures whom no matter where one turns to look at a serious discussion of economic theory in the period from the 1930s almost through the 1960s, Marschak is nearby.

GD: My recollection of Marschak is that he was indeed always present at staff meetings, seminars, that he had comments on virtually everything that could be understood to have interested him, and he was that way as long as I knew him. After New Haven, he eventually moved to Los Angeles. He was trained, I seem to recall, as an engineer before he turned to economics, and so his mathematical preparation must have been less complete than certainly Koopmans, Arrow, or myself. But he did not try to influence the direction of research in any direct way but maybe by his setting an example; he played an important role in the development of von Neumann-Morgenstern utility theory, and Herstein and Milnor, I believe, credit him with, what (reading the paper), yes, they have five references, and Jacob Marschak is one of them. They say "Marschak attacked the subject again." Then there is the mention of Herman Rubin, who was around. I barely knew him but he must have been very much around during the time of the development of the work on estimation. He was, as I understand it, a somewhat difficult person, but this is a secondhand piece of information.

ERW: There was little connection between the Harvard/MIT people, what became the group around Samuelson and Solow, and the older George Birkhoff and E. B. Wilson and so on. That set of people seemed to be entirely apart from the individuals and intellectual connections of the Cowles group. The Cowles group, though not self-contained, did not intersect with Samuelson and the arguments in *The Foundations of Economic Analysis*. There is a sense that mathematical economics was all of one piece, though in fact, you see that the pieces were quite disparate. From the perspective of being at Cowles was this a Chicago/Harvard division? Was it a difference in mathematical taste or problem taste or did the issues never come up? You yourself said you weren't aware of Samuelson's work until at least the conference in Salzburg. But the kind of research program that was suggested in the *Foundations* when it appeared as one piece doesn't seem to be very connected to the interests of Cowles. Is this fair or correct?

GD: No, that is true. That was quite the perception. There was no disagreement, hostility, of any kind at Cowles that I am aware of. There were indeed not many contacts to that group, and the work of Samuelson was very well known at Cowles as later was the work of Bob Solow. Where Cowles people met Samuelson and Solow was, I imagine, perhaps at RAND in the summer, but there were many fewer opportunities in those days for meetings of the American Economic Association. So my recollection of those contacts is that they were not numerous, not systematic. You are right that the mathematics are different in Samuelson's *Foundations*, and as for the problems themselves, there is of course a similarity of them to classical approaches to mechanics. But many of the messages that the *Foundations* contained did not influence the research at Cowles.

ERW: Is this to be expected? In your view was there simply a different kind of agenda? The one person who connects of course would be Larry Klein.

GD: Klein had left by the time I came to the Cowles Foundation. His interest was in econometric models mostly. Before I came to the U.S. I had read his book, *The Keynesian Revolution*, and that is the kind of problem that I never studied at the Cowles Foundation. I do not know how frequently it was discussed before my time there.

ERW: Was that more in your view having to do with the shift of Koopmans's interest from econometrics? You say he stopped working in those areas with the production of the volume on estimation.

GD: Of course. Koopmans and Marschak had turned to dynamic problems before I joined the Cowles Foundation but to my knowledge did not work on them after I was there.

ERW: Was there any discussion or interest in dynamics as such at Cowles at that time or were they simply non-issues?

GD: I think that I did not hear dynamics discussed either positively or negatively. I had my own views which probably I expressed, but I don't think it was discussed at seminars; indeed there was a remarkable shift of interest in the Cowles group. I would say "fortunately" because it was more to my taste, but before 1950 there was the attention to estimation of econometric models which became increasingly an econometric effort after 1950. And dynamic problems were not discussed after 1950.

ERW: Was there any sense of tension, conflict between the Chicago economics group itself and the Cowles people?

GD: Surely. And that must have been much more obvious in the department meetings which I did not attend. But I am sure when I say that tension occurred between, let us say, Milton Friedman and the Cowles group it must have been substantial from many different grounds. Because at Chicago the non-Cowles people were devotees of Alfred Marshall, and the Cowles group took a more general equilibrium viewpoint, and that was one difference. And I am sure that the non-Cowles group thought that the Cowles group used far too much mathematics. And then there were ideological differences. One of the issues of the day was rent control, and this found its way into our discussions. But occasionally antagonism flared up. Milton Friedman did not come to staff meetings, which were private affairs, but we had those seminars which were open to the public, and in those I would see Friedman, not regularly but often. That is the fundamental difference between staff meetings which were attended by small groups and seminars in which there may have been fifty more or less according to the speaker. So staff meetings were intended for the communication of research sometimes in progress. Seminars were more formal, sometimes with outside speakers, speaking about more finished research. Later, in the Yale set-up, there was a collection of houses and the Cowles Foundation was housed in one building and the department in another, etc. So the Cowles Foundation preserved its identity certainly at first. Later on, it grew and grew and I think it became very large after I had left but certainly in the first years after the move to New Haven it was not a large group.

ERW: In the earlier period at Chicago in which you're involved, one has more of a sense of the Cowles group being outsiders to the economics profession but the general success of the work, the acceptance of the work and by the time of the move to Yale, the individuals in the Cowles group were greatly sought after, were coming to be well respected within the profession and could no longer really be thought of as outsiders, may not have formed a dominant, strong center of the economics theoretical community as we now see it became, but were not outside the mainstream. Is that characterization appropriate?

GD: You're right. And I seem to find that Tobin was very much an insider and became director of the Cowles Foundation, that helped to change that. And as we saw, he was director for seven years.

ERW: Was part of the, what later I understood to be, antagonism between the Chicago department and the Yale department associated with the feeling that the Cowles people with whom there was beginning to be some tension, were now at Yale?

GD: Yes. I knew there was this tension but I was not acutely aware of it. And to some extent because I was not a member of the department at Chicago, when the decision to leave Chicago was taken up, it must have been a difficult decision. The Chicago department must have certainly after having been criticized, realized that it had something to lose, something important, and in that case to say that one group of persons did work that was not consonant to the work . . .

ERW: It wasn't economics.

GD: Yes, that's right. And later on, much, much later on I remember at the National Academy of Sciences I had a few contacts with Ted Schultz in particular and he was surprised to see that I had no antagonism whatever to the department of economics at Chicago. On the contrary. I realized it was a very good department. So that is in a way when emotions were much cooler and many years after, I became aware by small bits of things that there was tension but indeed when you were in Chicago, sometimes it was very obvious.

ERW: Especially over, as you mentioned, rent control, or whatever the issues were at the time.

GD: And sometimes that tension cleared up in just one or two sentences.

ERW: Are there any things that you would like to stress over this account in talking about the mathematization of economics over that period that you feel I should be pointed to?

GD: I can tell you my dominant feelings and that you must be aware of it. I was, before I joined the Cowles Foundation, in a group where mathematics was under some suspicion, even by mathematical economists. It went too far. So the concept that mathematical rigor was a sine qua non condition of theoretical work was something that I learned possibly in the book by von Neumann and Morgenstern and in a larger sense from the Cowles Commission in 1950 after I joined it. I felt entirely free and more or less I felt approbation, approval, of the work I wanted to do, which was very important. So I have an extremely good memory of those first years, the 1950s at the Cowles Foundation because it was an expansion of the possibility of working as I wished. And whereas before I was in a group which felt mathematics went too far and points of rigor were not terribly important, at Cowles I came to think, very quickly, that full understanding of a problem required no compromise whatsoever with rigor. There is that amusing story that is told by Charles Roos somewhere: he had returned to him a paper he had sent and it was turned down in turn by an economics journal and a mathematics journal, and each one of them said its fine, we'll accept it, just remove the other things. As Roos told that story I think that was one of the arguments in favor of founding *Econometrica*, which had as one of its principles that nothing could be rejected because it was too mathematical. And that was such a time, in the early 1950s it was possibly the only theory journal at that time that did that. Nowadays, a contributor has a choice; indeed I could name right away five or six which would not object per se to the use of sophisticated mathematics.

ERW: Now, at this point the marginalized group within the profession are the historians of economics and economic historians.

GD: Yes, and I have tried to react against that. I don't know if you noticed when I was president of the AEA, I had the exorbitant privilege of choosing the Ely Lecturer, and I chose David Landes.

5 Negotiating at the Boundary

with Ted Gayer

No matter how insulated is his laboratory or solitary his research, the scientist always operates as a social being in two fundamental respects. First, the language or symbolic mode of his conceptualizations—both its lexicon and syntax (that is tokens, chains, routes, and networks of his conceptual moves)—has necessarily been acquired and shaped, like any other language, through his social interactions in a particular verbal community, here the community of scientists in that discipline or field. Second, in the very process of exploring and assessing the 'rightness' or 'adequacy' of alternative models, the scientist too, like other professional evaluators, characteristically operates as a metonymic representative of the community for whom his product is designed and whose possible appropriation of it is part of the motive and reward of his own activity.

—Barbara Herrnstein Smith, *Contingencies of Value*

Mathematical economists often claim that one can translate between mathematics and economics. Paul Samuelson claimed in *Foundations of Economic Analysis*, following his mathematical mentor's mentor, J. Willard Gibbs, that "mathematics is a language." This belief in translation is often based on an implicit realist epistemology that suggests 1) the economy exists autonomously; 2) it can be represented by ordinary language propositions; and 3) the language of mathematics is useful in translating and operating with those propositions characterizing that autonomous existence. An implied corollary of this position is that any disagreement between an economist and a mathematician on the nature of a mathematical proof is due to a misunderstanding of the assumptions or the logical reasoning of the proof. And any disagreement on the economic implications of the mathematical proof is due to mistranslation or a lack of understanding of the underlying economic reality. Consequently there is an implicit "right way" to understand how economists and mathematicians can negotiate the more or less rigid boundary which separates their disciplines.

Nevertheless, a number of studies document communication failures between mathematicians and economists. The most prominent of such studies detail the failures of economists to comprehend what mathematicians are trying to tell them about their mathematical economics work. For example, Ingrao and Israel (1990) have discussed the problems Pareto had understanding the criticisms made of his work by Vito Volterra. Mirowski (1989, 243–48) has examined the failure of Leon Walras to make sense of the letters from Hermann Laurent, who had tried to ask Walras about the nature of the integrating factor in the equilibrium conditions for marginal utility: that discussion went nowhere and ended when Walras "started suggesting to others that Laurent was part of a plot against him" (245). In this chapter I shall instead explore the attempt of a mathematician to work within the economics community. The correspondence between the economist Don Patinkin and the mathematician Cecil Phipps exhibits the process by which members of these different disciplinary communities attempt to reconcile differences.[1] Within their correspondence, Patinkin and Phipps discuss the validity of a mathematical proof that emerged in Patinkin's economic research. Their correspondence sheds light on the complexity of achieving a common understanding about the role of assumptions, the nature of proof, and the meaning of mathematical modeling—issues that challenge the belief that economics can be translated into mathematics.

Introducing Don Patinkin

Don Patinkin was born in Chicago, Illinois, in 1922. In his posthumously published paper "The Training of an Economist," Patinkin recalled that before entering college his vocational aptitude results "showed a high aptitude for mathematics. But we were still living in the shadow of the Great Depression and everyone knew that mathematicians went hungry. So the advice to me was to become a statistician—with the explanation that a statistician was a mathematician who could make a living." Patinkin went on to receive his Bachelor's degree in 1943 (entering as a third-year student in 1941), his Master's degree in 1945, and his Ph.D. in 1947—all from the University of Chicago. He then held teaching positions from 1946 to 1948 at the University of Chicago, rising to the rank of Assistant Professor. After spending a year as an Associate Professor at the University of Illinois, he immigrated to Israel in 1949, and there spent the remainder of his career at the Hebrew University in Jerusalem, eventually becoming its president.

Because we will be concerned with understanding Patinkin's correspondence with a mathematician, it is useful to discuss the kind of mathematical training that Patinkin received as a graduate student at the University of Chicago. Patinkin's primary teacher of graduate economic theory was Oscar Lange, who taught systematic courses in microeconomics and macroeconomics.

> But Lange's most valuable course for me was the one on mathematical economics (i.e., what was then called mathematical economics!). Here he systematically took us through the mathematical appendix of Hicks's *Value and Capital* (1939), as well as Paul Samuelson's pathbreaking article on "the stability of equilibrium" (1941), subsequently reproduced as chapter nine of the *Foundations of Economic Analysis* (1947). My lecture notes from this course served me as a "reference volume" for many years to come. (371–72)

We thus see that Patinkin, taught by Lange, who was mathematically quite adept, was working through some fairly sophisticated material rather early in his graduate career. Moreover, Patinkin's graduate career roughly corresponded to the period in which the Cowles Commission took hold at Chicago. "In 1943, shortly after Jacob Marschak's arrival, Ted Anderson, Trygve Haavelmo, Leo Hurwicz, Lawrence Klein, Tjalling Koopmans, Herman Rubin, and (somewhat later) Kenneth Arrow had joined the staff—some of them with joint appointments in the department" (375). Thus Patinkin had contact with future Nobel laureates Klein, Koopmans, and Arrow, and was able to interact, on a regular basis, with a group of extremely mathematically sophisticated economists. Indeed, Patinkin received an SSRC Fellowship for 1946–47, which enabled him to serve as a junior member of the Cowles Commission while he wrote his doctoral thesis, and he spent the following year at Cowles as a research associate.

Patinkin was not a mathematician. Neither was he an applied mathematician. However, from the perspective of the discipline of economics at that time, Patinkin would have been regarded as a hotshot mathematical economist. After all, there were very few places in America where mathematics was even regarded as appropriate for students of economics: the University of Chicago stood out among other institutions in this regard even though, with Hotelling at Columbia, Samuelson moving to MIT, and Evans at Berkeley, it was possible to receive training in mathematical economics elsewhere. But Chicago was almost unique at that time in having a large and active group to defend those interests. In fact, when Koopmans succeeded Marschak as research director of Cowles, he began making overtures to the eminent Chicago mathematics department to establish a mathematical statistics unit. If one was interested in mathematical economics in the United States in the 1940s, the graduate program at the University of Chicago was the place to pursue this interest.

Patinkin received his mathematical instruction primarily from Marschak. Under Marschak he took "an advanced graduate course in mathematical economics devoted to solving the problems in the second half of R. G. D. Allen's *Mathematical Analysis for Economists* (1938)" (375). He was the only student in that class and thus, in effect, had a mathematics tutorial from Marschak. It was this connection that lead Patinkin to seek out Marschak to be chair of his doctoral thesis committee.

Patinkin's dissertation developed from a graduate student paper on "market-adjusting and inventory equations." With the encouragement of H. Gregg Lewis, in 1947 he began asking questions about the possibility of interpreting involuntary unemployment as labor being off its supply curve. Though his thesis committee consisted of Marschak (chairman), Lewis, Paul Douglas, and Theodore O. Yntema, only Marschak and Lewis were really involved in his dissertation as advisers (379); the former brought the strength of mathematical modeling, while the latter was developing his strengths in labor economics: "The thesis consisted of two parts: the first dealing with the mathematical consistency of a general-equilibrium system with money; and the second with unemployment as a manifestation of an inconsistent system. . . . I still remember my excitement when I thought of interpreting it [involuntary unemployment] in terms of an inconsistency in the system which prevented it from reaching an equilibrium position" (379).

As Patinkin notes, "practically all of the first part of my thesis appeared in two *Econometrica* articles 'Relative Prices, Say's Law, and the Demand for Money' (1948) and 'The Indeterminacy of Absolute Prices in Classical Monetary Theory' (1949)" (1). The 1948 article and the first three parts of the 1949 one were more or less unchanged from the thesis. "These first three parts were primarily devoted to demonstrating the invalidity of the traditional dichotomy of general equilibrium theory between the determination of equilibrium relative prices, on the one hand, and the equilibrium absolute level of prices on the other. . . . However, the last ten paragraphs of this part of the article—which I then termed a 'modified classical system,' in which there was no such dichotomy, but in which the classical neutrality of money a la quantity theory nevertheless held—did not appear in the original thesis" (380–81). Patinkin recalled that those ten paragraphs were, for him, the heart of the argument: "I still have vivid memories of the moment of truth when everything suddenly fell in place: when after long being troubled by the problem, I suddenly realized that the economically meaningful way for the commodity demand equations to depend on the absolute price level (and thus to avoid the invalid dichotomy) without violating the neutrality of money was to have them depend on the real value of money balances" (381).

Patinkin was later to reflect that "like most doctoral students (then, and I'm afraid even more so now), I attributed too much importance to technique and formal mathematical analysis. And so my thesis gave much emphasis to the rigorous derivation of theorems from definitions, assumptions, and preliminary lemmas, while devoting inadequate attention to the economic interpretation of the analysis" (383). Yet he notes that the last ten paragraphs of the 1949 article were different in that they contained an economic interpretation of the mathematical results. Consequently, the heart of that 1949 *Econometrica* article, those ten paragraphs, set the stage for Patinkin's own reevaluation of the interconnections between mathematics and economics: "In the year following my 1949 article, I gradually developed the philosophy that the mathematical analysis of any economic problem is not complete until it is given an intuitively appealing economic interpretation. From experience over the years, I also learned that when there was a contradiction between the mathematics and the intuition, it is not always the intuition that was at fault, but frequently an implicit (and sometimes explicit) misguided assumption in the mathematics. Thus resorting to intuition as well as mathematics provides the most useful check on the analysis. It is a way of carrying out a fruitful dialogue with one's self. And it is the dialogue that I later carried out between the text and mathematical appendix of *Money, Interest, and Prices*" (383).

While writing his two *Econometrica* articles that set out the major results of his thesis, Patinkin became committed to the new state of Israel and a career away from the United States. Eventually he would become widely known among economists for his work on the neoclassical synthesis, which reached its pinnacle with his 1956 book *Money, Interest, and Prices*. But in the 1940s he was a young and very confident economist. He had just been trained and educated at the intellectual center of the American economics community, the American mathematical economics community, and he was looking to make his mark.

Introducing Cecil G. Phipps

The 1955 edition of *American Men of Science* has telegraphic biographical information on Cecil Glenn Phipps. He was born in Skidmore, Missouri on 24 July 24 1895, received his Bachelor's degree in mathematics from the University of Montana in 1921, and was an Instructor and Assistant in Mathematics at the University of Minnesota from 1921 to 1924, where he received his Master's degree in 1924. Phipps then went to the University of Florida as an Instructor from 1924 to 1927, going on leave to return to Minnesota to receive his Ph.D. degree in 1928, returning thereafter to Florida with the rank of Assistant Professor. Phipps's doctoral dissertation concerned "problems in approximation by functions of given continuity." This exercise in mathematical analysis, approximation theory, examined how an arbitrary function could be approximated by a function of N-times continuous differentiability. He obtained the rank of Associate Professor in 1929 and Professor in 1943. In that 1955 *American Men of Science* volume he lists his areas of interest as "approximation of functions of real variables" and "foundations of mathematical economics." Phipps left Florida for Tennessee Tech in 1960.[2]

The notice of the April 1933 meeting of the Mathematical Association of America, Southeastern Section, includes a paper by Phipps on "Subfreshmen Mathematics." During the same period, the catalog of the University of Florida showed that Phipps "is teaching plain trigonometry and solid geometry, elementary mathematical analysis, integral calculus, and advanced topics in calculus." Later, on the eve of World War II, we find Phipps teaching a general science course, "Man and the Physical World." Additional material on Phipps comes from Paul Ehrlich's history of the University of Florida mathematics department. The first mention of Phipps occurs in chapter 5 of this history. Ehrlich notes with some surprise that "a check of the *Mathematical Reviews* author index for 1940–1959 reveals that Cecil G. Phipps's research area was mathematical economics. It is interesting to note that a paper Phipps published in 1952 on 'Money in the Utility Function' received a *Mathematics Reviews* report by the eminent mathematical economist Kenneth Arrow." In chapter 8 of his manuscript, Ehrlich mentions that Robert George Blake received a Master's degree in May 1945 under Phipps's supervision. The Master's thesis was titled "Circular Arrangement" and was completed in conjunction with what was referred to as the Elasticity Theory Group. We also have a reference, among a group of recollections of elderly Gainesville residents in that history, that a Mrs. Pirenian grew up in Gainesville, and "as a Gainesville high school junior, she valued the privilege of studying geometry under Mrs. Dorothy Phipps, wife of Professor Cecil Phipps of the University of Florida mathematics department."

In an interview Professor Franklin W. Kokomoor gave to the Florida Oral History Project, concerning his joining the University of Florida mathematics department in 1927, he recalled that "when I first came as I said in my notes that I gave you there, there were only four of us in the department and three of us were new. Two others [one of course was Phipps] besides myself, just new. We taught 15, 18 hours a week of class work, and that was the full contents of our mathematics offering here. But in the course of time, as we got more students and new colleges, and new colleges needed new mathematical services and so on, we kept on growing until I retired. . . . The student enrollment in 1927 was just over 2,000, so you see we were a small and close-knit school. The professors developed strong ties of loyalty and pride in their university, perhaps because each of us became intimately involved in campus activities, as well as teaching." Apparently, the starting salary for Phipps as an assistant professor in 1927 would have been the same as Kokomoor's, $2,500 a year.

In Professor Charles Crow's manuscript on the early history of the University of Florida (drawn on by Ehrlich) there is a reference to a monthly faculty discussion club, the Atheneum, already established by the time of the Sledd presidency in Lake City, prior to the move of the campus to Gainesville: "Members would work up lectures on subjects outside their academic specialties for presentation to the others at these monthly meetings. Professor Samuel Gould Sadler informed me that Professors Franklin Kokomoor and Cecil Phipps had been participants in this Atheneum Club; Phipps recruited Sadler into membership." Phipps's interests thus extended beyond mathematics, though they may have been more engaged the closer the subjects were to mathematics, a hardly surprising observation. Yet as we will see later in this chapter, Phipps frequently referred, in his letters, to his membership in a group of scholars who read economics papers as critics. In William Gilbert Miller's Ph.D. 1951 dissertation, which we believe is the only dissertation supervised by Phipps, there are frequent references to "Unpublished Notes on Econometrics," which consists of Phipps's "lectures and papers on the subject, . . . with contributions being made by Dr. G. B. Lang, Dr. M. D. Anderson, Mr. Ernest Lytle, Mr. R. N. Conkling, Mr. H. E. Whitsett, and the author" (iii). This group appears to have been drawn from individuals—faculty and graduate students—who had interests in econometrics and statistics.[3] Some of the individuals may have had cross-connections with Phipps from his involvement in the Atheneum club.

Miller's dissertation, "The Mathematics of Production and Consumption in a Static Economy," reflects many of Phipps's views, and even his words, as we shall see when we come to examine Phipps's correspondence. In the preface Miller writes of the "errors and misconceptions" that occur in the new science of mathematical analysis as applied to economics. "Nowhere in the literature have I been able to find a complete and correct mathematical treatment of the general case of production and consumption" (ii). Miller states that "mathematics when correctly used provides the economist with a powerful and versatile tool, but the rules governing its use are rigid. If the economist is to support his argument with mathematical methods, he must adhere strictly to the mathematical rules" (ii). As we will see, this belief in the sanctity of mathematics, and the misuses of mathematics by economists, was certainly held by Phipps.[4]

Phipps's interest in mathematical economics seems to have taken hold in the late 1940s and it continued through the 1950s. He appears not to have published anything in mathematics in that period, and with the appointment of an outsider, John Maxfield, as the new department chair charged to make appointments of young research faculty, Phipps became increasingly marginalized within the department. Memories of older faculty confirm Phipps's displeasure over the Maxfield appointment; indeed he resigned in protest over it in 1960 and accepted the invitation of a former Florida graduate student, Ralph C. Boles, who at that time was chair of the mathematics department at Tennessee Technological University, to join its faculty. He taught at Tennessee Tech for about five years, conducting mostly upper-level courses and Master's-level courses, primarily in mathematical analysis.

A former colleague at Tennessee Tech, Reginald Mazeres, recalled Phipps quite clearly (1998, personal communication). "It is obvious that at one time he'd been a brilliant man." However, "Phipps was not an easy man to get along with."[5] He was a good conversationalist and extremely opinionated. In the Tennessee Tech years, he had problems with both people and mathematics: "Some of his work was offbeat." Mazeres recalled that at a MAA regional meeting at Emory University in 1966, Phipps "presented a paper on the empty set." It was there, in the discussion, that "some former colleagues of Phipps, University of Florida mathematicians, gave him a rough time . . . they were extremely rude." Phipps apparently had an obsession with the empty set, and would object in departmental seminars at Tennessee Tech when a mathematician wrote down the symbol for the empty set. He argued that since the empty set had no elements, it could not be characterized, and therefore it could not be used: since no one had any positive idea of nothingness, thus nothing itself could not be characterized. Putting the most positive possible spin on this lunacy, his then chair told us that Phipps argued that 'if a set is empty, it has no elements and thus there's nothing to say about it'" (Boles 1998, personal communication).

Phipps in Economics

As background to the Patinkin-Phipps correspondence, we must understand that Cecil Phipps focused much of his attention on reviewing applied mathematical work. As already noted, he was a member of a group of mathematicians and statisticians who sought to make a thorough examination of the existing literature of applied mathematics and to point out occurrences of unsound mathematics. In a letter to Don Patinkin dated 20 December 1949 Phipps wrote: "I examine [applied work] for soundness of the mathematics in them. If it is faulty, the article is worthless until the defect is corrected. When I read an article or a book whose results are based on mathematical deductions, I expect the same quality of mathematics as I would expect in a master's or doctor's thesis."

Phipps's emphasis on examining work in applied mathematics led him to publish three notes, between 1950 and 1952, on what he saw as mathematical inaccuracies in articles by Don Patinkin, Gerhard Tintner, and Milton Friedman.[6] On 16 February 1950 he expressed his disgust with these papers to the editor of *Econometrica*, William Simpson:

> My feelings on this matter are intense; some might say, bitter. I am shocked at the misuse and abuse of the prestige of mathematics in reaching conclusions not logically justified by the given assumptions, especially when the "proof" is demonstrably false. It would save embarrassment to everyone (and printing costs) if every article in *Econometrica* were to receive before publication the same scrutiny which I am giving these articles after publication.

Phipps's first foray into economics was his one-page "Note on Tintner's 'Homogeneous Systems'" (1950a). In it Phipps corrected Theorem 3 of Tintner (1948), which claimed the following: "A function f is assumed to be homogeneous of zero degree in the variables $u_1, u_2, \ldots, u_m$. These variables are themselves functions of the M variables $v_1, v_2, \ldots, v_m$. The function f remains homogeneous of zero degree in the new variables $v_1, v_2, \ldots, v_m$ if the old variables $u_1, u_2, \ldots, u_m$ are homogeneous functions of the same arbitrary degree in the variables $v_1, v_2, \ldots, v_m$."

Phipps correctly points out that the conditions for this theorem are sufficient, but they are not necessary. He uses a simple example to demonstrate his point: "Let $f(x,a) = g(x) [a_2 - a_1] / a_3$. . . . Then f is homogeneous of degree zero in the a. Next let $a_1 = c_1 + 5$, $a_2 = c_2 + 5$, and $a_3 = c_3$. After this substitution, $f(x,c)$ will be homogeneous of zero degree in the c although the a were not homogeneous of any degree in the c."

Phipps's concern with Patinkin, which we will discuss in detail below, dates from the time he published a note on Patinkin in *Econometrica* in 1950 (1950b). As is usual practice for writings by mathematicians, that *Econometrica* paper was itself reviewed in the *Mathematical Reviews*, volume 11. The reviewer, M. P. Stoltz wrote, "several deductions are drawn from the joint assumptions of perfect competition and utility maximization as a behavior rule. . . . It is now argued that Patinkin . . . finds the classical system inconsistent because of contradictory assumptions. . . . *The author's criticism seems invalid to the reviewer*" (Stoltz 1951, 530; emphasis added).

Phipps's paper, "Money in the Utility Function" (1952a) concerned the Patinkin themes in more detail, and it also received a notice in the *Mathematical Reviews*, a notice that Ehrlich had thought surprising. The reviewer, Kenneth Arrow (!), pointed out that Phipps made "the utility function depend on the following variables: the rates of consumption of commodities (the usual argument); the amounts of each commodity which could be purchased with a stock of Money; the amounts of each commodity which could be purchased with a stock of invested Savings; and the stocks of commodities held by the individual" (Arrow 1954, 366).

In that same year Phipps (1952b) criticized the mathematical reasoning used in Friedman (1952a). In that article, Friedman set out to demonstrate "that an alleged 'proof' of the superiority of the income tax [over the excise tax] is no proof at all, though it has repeatedly been cited as one." He presents two goods, X and Y, and then assumes an excise tax (entirely shifted to the consumer) of 50 percent placed on good X. This tax (called Excise Tax A) results in a rotation of the budget constraint and a new equilibrium consumption bundle. He then supposes that, instead of the excise tax, an income tax (called Income Tax A) is imposed to yield the same revenue. This shifts the original budget constraint in, where it crosses the equilibrium bundle of the excise tax. The budget constraint resulting from the Income Tax A offers a higher level of utility than the budget constraint resulting from the Excise Tax A, and thus the income tax is preferable.

To demonstrate the alleged fallacy of this proof, Friedman assumes that Excise Tax A is already in effect. He then assumes that a second excise tax of 50 percent (called Excise Tax B) is placed on good Y. Using the same reasoning as before, he demonstrates that placing an income tax that yields the same revenue (called Income Tax B) is preferred to the excise tax. Therefore, "Income Tax B plus Excise Tax A is preferable to Excise Tax B plus Excise Tax A." Since placing a 50 percent excise tax on good X and good Y is the same as a 50 percent income tax he argues, "Income Tax B plus Excise Tax A is preferable to Income Tax B plus Income Tax A." However, "when Income Tax B is removed from both sides," the result is that "Excise Tax A is preferable to Income Tax A." The contradiction of the original proof "follows rigorously" from the same argument. In his own note, Phipps contests Friedman's paper by pointing out that it is incorrect to drop Income Tax B from both sides. The relative price of good X is higher when Excise Tax A is in effect than when Income Tax A is in effect. Therefore, the income tax removed from the left-hand side is a lighter tax than the income tax removed from the right-hand side.

Friedman's "A Reply" (1952b) immediately followed Phipps's paper in that issue of the *Journal of Political Economy*. In it, Friedman began by saying, "Professor Phipps is entirely correct that my attempted reductio ad absurdum of the usual excise-tax-income-tax argument is a dud" (334). Nevertheless, he goes on to say that "Phipps is entirely wrong in supposing that this inexcusable blunder affects the validity of the rest of my paper . . . yet it [that particular argument] is only a flourish, not an essential part of the analysis, and it can be deleted bodily with only minor verbal changes in the rest of the paper" (335).

From the public record we seem to have a straightforward case of an individual, who happened to be a mathematician, writing some short comments and corrections to some papers written by economists on quite diverse subjects. Though he seemed to be writing on minor points, not directly related to the papers' real contributions, Phipps seemed to have a respectable engagement with the mathematical economics literature, at least as a close reader of the written text. Just as one of my colleagues assigns graduate students in his International Trade course to find minor errors in published papers, and to pen "A Comment on X" for part of the course grade, so too Phipps appears to have a self-imposed assignment to write comments on mathematical errors in economics papers.

Mathematical Economics at Midcentury

The presence of mathematics in economics is not a recent phenomenon. In justifying its use, economists have frequently described mathematics as a tool that facilitates the analysis and exposition of an economic problem. According to this much-held view, mathematics offers another way of stating an economic theory, and one can (and should) clearly translate between the mathematical model and the economic model. As early as 1838 we see Augustin Cournot defending his work by remarking that "the importance of mathematical symbols is perfectly natural when the relations between magnitudes are under discussion; and even if they are not rigorously necessary, it would hardly be reasonable to reject them, because they are not equally familiar to all readers and because they have sometimes been wrongly used, if they are able to facilitate the exposition of problems, to render it more concise, to open the way to more extensive developments, and to avoid the attractions of vague argumentation" (Cournot 1963, 2).

By the late nineteenth century the linking of mathematics to economics was seen as a means of establishing economics as a science. Alfred Marshall's long war of attrition to establish an Economics Tripos in the Cambridge system is a testament to this connection, yet I have argued in chapter 1 that Marshall, eager for mathematical connection and a former Second Wrangler himself at Cambridge, was increasingly disconnected from the world of mathematics as it was developing outside late Victorian England. We can thus read his letter to Bowley, discussed earlier, as an extended gloss on the workings of the translation metaphor.

The issues were different by the middle of the twentieth century, when the correspondence between Patinkin and Phipps takes place. By this period, Marshall's translation metaphor takes root in Samuelson's view that "mathematics is a language." But Samuelson's view is not an isolated one. The translation metaphor is also a central theme of the famous Tjalling Koopmans (1957) essay "On the Interaction of Tools and Reasoning in Economics" where he presents mathematics as a tool to be used to solve, or facilitate the solving of, economic problems that are distinct from the mathematical tools themselves:

> While "problems" are to some extent posed by conditions and needs of society, "tools" and states of training in the use of tools are part of

> the personal acquaintance of the investigator. It is true that it is already difficult, at best, to be objective about what are valid answers to a given social or economic problem. . . . It is even harder to be objective about what are promising tools for unsolved problems: the usefulness of our own individual minds and of the investments of personal effort sunk in our training and direction of interest are involved. (170)

Koopmans here employs a powerful and often compelling image of mathematics, where it is one of several tools he considers (among statistics, computer programs, etc.): mathematics is a distinct body of knowledge to be employed to solve problems in a separate body of knowledge. As we saw in chapter 4, this paper was considered by Gerard Debreu to be one of the compelling presentations of the Bourbakist image of mathematics for economists as well. Just as a hammer is not a building, but can be employed to build one, neither is mathematics economics, though it can be used to construct an economic argument. Better hammers, and knowledge of how they can be used, may allow better structures to be erected.

The belief in the translation of mathematics is often accompanied by a view that the language of mathematics is useful in describing an autonomous economic reality. This view surfaces in Koopmans's characterization of "economic theory as a sequence of models."

> Each model is defined by a set of postulates, of which the implications are developed to the extent deemed worthwhile in relation to the aspects of reality expressed by the postulates. The study of the simpler models is protected from the reproach of unreality by the consideration that these models may be prototypes of more realistic but also more complicated, subsequent models. The card file of successfully completed pieces of reasoning represented by these models can then be looked upon as the logical core of economics, as the depository of available economic theory. (142)

Koopmans would have us study the mathematical model in lieu of the economic substrate directly, perhaps because the model allows difficulties to be separated out, and addressed seriatim in increasing complexity.[7] The image, for that is what it is, is that of a map of a region as a representation of the region different from that region, a picture as a representation of an object different from the object. A mathematical model is a metaphor for the economic problem under investigation, an X which is considered as a Y. In mathematical terms (!), it is the case that the structures of X and the connections in X—the mappings associated with X—are themselves mapped consistently on Y and Y's mappings. X and Y are homeomorphic: X is relevantly like Y. It is not, however, easy to make sense of "relevantly like."

At about the same time that Koopmans was writing his essay, another future Nobel laureate was setting out his views on the topic. In a series of lectures that George Stigler delivered at the London School of Economics in 1949 (Stigler 1969 [1949]), the fourth—"The Mathematical Method in Economics"—provides an interesting window on the role of mathematical economics at midcentury. Stigler was not, of course, trained as a mathematician, but he had reasonably sophisticated knowledge of at least quantitative methods in what would come to be called applied economics/econometrics, and his view of mathematics is one that was widely shared at the time of his writing. "Because mathematics is the premier language of logic, it is a method: a method of drawing exact deductions from given premises, and of verifying the logical consistency and adequacy of the premises. It follows that mathematical economics is a thing without content" (37). Stigler notes in a footnote that "the laymen's appreciation of mathematics must necessarily be based upon authoritative hearsay; I have found R. Courant and H. Robins, *What is Mathematics?* especially illuminating" (37). This idea that mathematics itself is contentless expresses much the same idea as mathematics is a language, not a set of referents.

Stigler sets as his task his "wish to explore special claims for the mathematical method, claims that transcend its admitted power and usefulness" (38). He argues that it does not necessarily lead to good economic theory, nor does mathematical exposition have an inherent clarity. Stigler rejects the claim that "the mathematically trained economist states his concepts more clearly, on average, than the unmathematical economist" (40), although he concedes the claim that "in certain types of [economic] analysis, the mathematical method is indispensable" (40).

On balance, Stigler wishes to argue that a mathematical structure of economic analysis must be based upon an analysis of uniformities in the subject matter:

> In the present early stage of economic study, the economist as scientist must be largely occupied with the isolation of these uniformities in his subject matter . . . until we possess many uniformities, we cannot erect broad analytical systems which are likely to be illuminating in the area where uniformities have not yet been isolated. This is true because it is a variety of uniformities calling for systematization that gives rise to a useful analytical system. (41)

We see here again an almost naive realist view of science, one that had been problematized long ago by individuals like Fleck, Bachelard, Duhem, and Popper: the idea of a preexisting economic world that could be described independently of any description, identified independently of any identification, represented directly and clearly independently of any representation, held real sway among most economists (and perhaps it still does!). On this view, economies "are," and economics "is," and mathematics is a non-natural mode of analyzing "it." Stigler's solution to these problems involves first that economists study more mathematics (though this presents the problem of substituting at the margin, as it is the study of economics that must be substituted against). But second, "in his publications the mathematical economist can provide along with his equations a *translation* of his results into words" (44, emphasis added). His penultimate paragraph sounds the call: "The mathematical economist can, if he wills, always meet this obligation. Even when the details of the proof must be shrouded in a fog impenetrable to the non-mathematical economist, the assumptions and the conclusions can always be stated clearly in the language of words, and heuristic derivation of the conclusions is probably always possible. The failure to provide these *translations* is a renunciation of the canons of scholarship . . . the queen of the sciences should not be made a puppet of a scientific oligarchy" (45; emphasis added).

Though Stigler in this essay paused at one point to chastise Samuelson, Stigler's view of mathematics is fully consistent with Samuelson's approval of the idea that "mathematics is a language." With this metaphor, Stigler is right to call for translation, for it is as if the economist of mathematical inclination is simply writing in Russian, or Chinese, or Martian. (Of course this discussion begs the entire question of whether, if mathematics were like Russian, it could then be translated, for there is a huge literature on the nature and meaning of translation difficulties, and how there are always incommensurabilities, and often major losses, in language translation.)

In any event, we see that many of those involved in mathematical economics during the middle of the twentieth century viewed mathematics as a tool for describing an autonomous economic reality: they believed that even as mathematics clarifies economic reasoning, it nonetheless is a specialized language that can (and many say, should) be translated into ordinary language.

In opposition to this view is our belief that mathematics is a separate and distinct set of discursive practices and arguments, in which Kuhn's incommensurability problem occurs in spades, and translation fails necessarily.[8] The correspondence between Patinkin and Phipps offers a case study of communication between an economist and a mathematician. Of course we do not claim that Phipps is perfectly representative of the mathematical community at midcentury. Instead, we view the correspondence as an instance in which the translation metaphor is challenged in practice: at issue thus were such questions as "How are mathematical proofs evaluated in economics?," "What is the value of rigor in argument?," and "What is the proper role of mathematics in economics?"

The Patinkin-Phipps Correspondence

The correspondence between Patinkin and Phipps concerns Patinkin's "Relative Prices, Say's Law, and the Demand for Money" (1948) and Phipps's objections contained in "A Note on Patinkin's 'Relative Prices'" (1950b). Patinkin was informed of the "Note" before its publication and on 9 December 1949 wrote Phipps to request a copy "in order to determine if a reply to your comments is in order." Thus began a protracted, and sometimes heated, correspondence on the validity of the mathematical proof used by Patinkin.

Patinkin examined the generalized Walras-Pareto classical model that considered a monetary economy. He claimed that, by assuming people do not derive utility from holding money, the classical model is consistent only if there are no stocks of money. This implies that the classical system does not determine absolute prices. Patinkin went on to argue that introducing money into the utility function represents the satisfaction derived by individuals from holding money as a means of dealing with uncertainty. The implications of assuming money is in the utility function are that it is impossible that all the demand functions should depend only on relative prices, and it is impossible for Say's law to hold.

Phipps's criticism (1950b) focused on Patinkin's proof that the classical system is consistent only if there are no stocks of money. After establishing the equilibrium conditions for the classical pure exchange economy, Patinkin then considered the implications of extending the classical system to a monetary economy. The classical system holds that people do not receive utility from holding paper money, so an individual will not plan to hold any money at the end of the period. Thus the budget constraint of the classical system must be restated as

$$\sum_{j=1}^{n-1} p_j Z_{ja} = \sum_{i=1}^{n} p_i \bar{Z}_{ia},$$

where the left-hand side represents the product of the consumption flow of the goods and the prices of these goods, summed over the n-1 goods. Good n is paper money, which is not included in the left-hand side of the budget constraint since the classical system assumes that people do not hold any money at the end of the period. The right-hand side represents the product of the initial stocks of the goods and the prices of these goods, summed over the n initial stocks. The summation on the right-hand side runs up to n, since there is no reason why the individual's initial stocks should not include money. Patinkin assigns the nth good, paper money, as the numeraire (i.e., $p_n = 1$). According to Patinkin, "This modification of the budget restraint is the crux of the entire argument. Consequently, though it is intuitively quite obvious, I shall prove it rigorously" (141). Patinkin derives the first order conditions and imposes the equilibrium condition that the excess demand functions (the difference between the consumption flow of a good and the initial stock of the good, summed over all individuals) must be zero. He claims that the only way this equilibrium condition can hold (i.e., the classical system is consistent) is if the initial stock of money for each individual is zero.

This inconsistency, wrote Phipps, "is obtained by his use of a set of assumptions which can readily be shown to be contradictory" (25). Phipps claimed that Patinkin's contradictory assumptions are the following: 1) people derive no utility from paper money and therefore money does not enter the utility function; and 2) the price of money relative to useful goods is one. "Using these contradictory assumptions," claimed Phipps, "Patinkin finds it quite easy to show the system under discussion to be inconsistent" (26).

On 30 December 1949 Patinkin wrote to Phipps defending his assumptions, arguing that they were not mutually contradictory.[9] His defense of his proof employs a traditional line of argument in monetary economics:

> Perhaps I can make this clear by a simple counterexample drawn from history: in England, for a long period, the guinea existed as a unit of account, without there being any in circulation. Now, clearly the price of a guinea was one (guinea); yet no-one held any quantities of guineas and nobody wanted to. The fallacy in your proof is that you do not take account of such a commodity: that is, a commodity in which there are not stocks in existence. If you will consider such a commodity, you will see that you cannot carry through the line of your argument.

On 9 January 1950 Phipps responded with a slight modification of his original argument. He now considered the possibility that there can be a good with positive price for which an individual has no demand or supply. However, he says, "This example is slightly different from your paper money. In your [classical monetary system] none (no single one) has a demand for money; it is a closed economy with [demand for money] equal to zero for all [individuals]. How then can one 'sell' his money if he has any?" Phipps also claims that Patinkin's English guinea analogy is inaccurate. He wrote, "We have the same thing here in the USA; our eagle is a $10-goldpiece which is not allowed to circulate. It has its equivalent in ten silver dollars which do circulate and can be used to buy food, clothing and shelter. The same remark applies to the guinea; 21 silver shillings, its equivalent, would buy, or could be exchanged for, a desirable good."

On 20 February 1950 Patinkin defended his guinea example. He wrote:

> I still insist that the example of the guinea is a perfectly good one. The fact that a guinea is defined as 21 shillings affects only the determinacy of the system; it has nothing to do with the consistency. I think that

you are being misled here by the word "equivalent." You certainly do not mean that a guinea has a value of one because it is "equivalent" to—in the sense that it can be exchanged for—21 shillings. A guinea does not exist; hence, it cannot be exchanged for anything.

Economists can sympathize with the frustrated Patinkin. The idea that a numeraire can be fictitious, like "the price level," or the price of a unicorn, is not a difficult idea, it is just one not grounded in ordinary language. Phipps may not like the game Patinkin is playing, but it is hardly inconsistent, let alone incoherent.

For that reason, it is a clearly annoyed Patinkin who wrote, on 16 February 1950, questioning Phipps's ability to examine the mathematical soundness of economic work. He suggests that economics is not equivalent to mathematical reasoning since economics cannot readily be translated into mathematical form. One needs a background in economics before one can appeal to mathematics:

> Your general objective of surveying mathematical economic literature from a rigorous viewpoint is very commendable. However, to accomplish this objective it is necessary to understand the purposes for which the economist is using the mathematical analysis. This should not be misinterpreted as saying that economic reasoning will make incorrect mathematics correct. But it definitely is intended to imply that failure to understand the economic background of the problem under discussion may well lead (as it has led in the three papers you sent me) to *mathematical* errors. I should imagine that a pure mathematician would hesitate to pronounce definitive judgment on a question in mathematical physics without first thoroughly investigating the physical conditions involved; I think the same should be true for mathematical economics. To reverse your dictum, writers on economic questions should be held responsible for an understanding of economic analysis (emphasis in original).

On 9 March 1950 Phipps defended the right of mathematicians to examine work in applied fields. He held that economics can be translated into mathematical language. One can argue about correct translation, but not about sound mathematical reasoning.

If we take any economic question as an example, the problem raised by this question translates into certain mathematical language. Sometimes there may be a difference as to the exact translation; in that case, one should consider the alternate possibilities. Once the problem is posed in mathematical language, it should be possible to reach definite conclusions or sets of solutions. In economics and in mathematical physics, the mathematician is prepared to say *definitively*, "if these are the initial equations, then this is the final result." (emphasis in original)

Thus, according to Phipps, economic background is not important except as it serves to help in the translation to and from mathematics. He admitted that "it is easily possible, of course, for one to describe in words one problem and, in effect, to translate another. This error occurs much too frequently in all fields of applied mathematics."

On 18 March 1950 Patinkin persists in his claim that one must first have a firm understanding of economics before engaging in mathematical reasoning on economic issues. "I quite agree with you that it is shocking that any published paper should contain mathematical errors. But I think there is something even more important than that. I think it is fundamental that any paper which has any pretenses of being a serious scientific work should be free of misinterpretations, misunderstandings, and misrepresentations. This is especially true of a paper which attempts to criticize the works of others. If the paper makes it patently obvious that the critic has made no serious attempt to read and understand carefully the text which he purports to criticize, then this is [the] most heinous crime that can occur in the intellectual world."

Patinkin then tried to turn Phipps's beliefs about the nature of mathematics against him.

In your last letter to me you state quite correctly (and how you ever interpreted me as saying anything to the contrary is a mystery to me) that the mathematician should only take a given system of equations and determine its solutions; it is not his job to go into the background of these equations in the non-mathematical fields from which they originate. But immediately afterwards, when put to the test, you reject the criteria you yourself have just set up. [The classical monetary system I set up] is a mathematical system; where it came from is no business of the mathematician; the only question he is called upon to answer it: Is this system necessarily inconsistent?

On 25 March 1950 Phipps attempted to find middle ground by agreeing that a critic should make a serious attempt at understanding the text. Nonetheless, he claimed that Patinkin does "not leave room for the possibility that the author may not have been clear, for various reasons. Hence, a paper may not be understandable; or it may possess several interpretations; or again it may contain a flaw overlooked by the author." With regard to Patinkin's argument on the classical monetary system, Phipps wrote: "I can only repeat the argument of my note and conclude that the price of this good [paper money], *relative to goods which still have marginal utility*, is zero!!!! On this point I am willing to stake the whole argument. Do you still maintain that the stated conclusion does not follow from the stated assumptions? Or do we differ as to what those assumptions really are? This point seems to be the actual basis for our differences. Until it is settled, neither of us will accept the criticism of the other" (emphasis in original).

In his reply on 12 April 1950 Patinkin again asserted that an understanding of economics is necessary before one can adequately criticize economic work. "I am firmly convinced that you and your group must spend at least one or two years learning the basic fundamentals of mathematical economics before any worthwhile criticism will be forthcoming." In this letter, he again addressed Phipps's claim about the contradictory assumptions, this time by turning the argument on its head. While Phipps argued that Patinkin's assumptions are contradictory and therefore lead to a proof of inconsistency, Patinkin countered that under his assumptions it is possible for the system to be consistent; thus the assumptions must not be contradictory. He claimed that the point of his argument is that the system is consistent only when money stock is equal to zero.

> You claim that the assumptions of [the monetary classical system] are inconsistent. What does this mean? It means that it is *impossible* to derive a system from these assumptions which will be consistent. Thus in order to find out if *my* assumptions . . . are inconsistent we must take *my* assumptions and see if they lead to a system which is *necessarily* inconsistent. . . . Now, what you must do in order to make your point is

to show that this system *under my assumptions*—specifically, under the assumption that [the price of money is 1], must always be inconsistent. For this is the meaning of your assertion that *my assumptions* are inconsistent. But you cannot demonstrate this; for as the theorem . . . explicitly points out, if [money stock is zero], the system *might* be consistent. Hence your argument falls to the ground. (emphasis in original)

On 1 May 1950 Phipps defended his assessment of Patinkin's assumptions with a discussion of 1) the consistency or inconsistency of a set of postulates, and 2) the dependence or independence of a set of postulates.

The dependence or independence of a set of postulates can, in theory at least, be demonstrated logically. If any postulate of the set can be derived as a theorem from others of the set, it is dependent upon them and should not logically be classified as a postulate. On the other hand, if a situation can be created whose conditions satisfy all postulates but one, or which contradicts only one, then that one is shown to be independent of the others.

The consistency or inconsistency of this set is not so easily dealt with. In the first place, it is *never* logically possible to *prove* the consistency of a set of postulates. *It is therefore useless to assume the consistency of the set* . . . since it is never possible to know for certain if this postulate is itself consistent with the others. Assuming consistency does not make it so. The best that anyone can do is to avoid any *demonstrable inconsistency*.

The matter of inconsistency is more definite. If it can be shown that a group of one or more postulates *requires* (not just permits as mentioned before) a certain conclusion which is contradicted, in whole or in part, by another postulate, then that other postulate is inconsistent with those of the group. In other words, the set is *demonstrably inconsistent*.

In this latter category, I place your assumption . . . that [the price of money] is one. (emphasis in original)

Phipps also claimed that Patinkin was incorrect in asserting that one must show that the set of assumptions must always lead to inconsistency.

"My statement implies much less; it implies that *some* conclusions drawn from this set of assumptions *may* fail to hold or *may* be inconsistent with other conclusions. . . . My statement admits the possibility of the given condition (inconsistency); your statement makes the condition mandatory. My statement allows the alternate possibility of deriving from these assumptions a system of conclusions which among themselves are consistent; your statement does not. The stronger statement of yours has no logical foundation. . . . My argument does not fall since it admits the possibility of a *limited* consistent set of conclusions derivable from these inconsistent assumptions. My argument states that, since your assumptions are demonstrably inconsistent, there is the ever-present danger that a derived conclusion will be found inconsistent either with certain of the original assumptions or with other derived conclusions" (emphasis in original).

On 1 June 1950 Patinkin suggested that the discussion was leading nowhere. He wrote, "I really do not think there is any point carrying on the discussion any further on this issue. I have made myself as clear as I possibly can be in my last letter to you, and there is nothing I can add to it." He does make one last attempt at explaining their differences.

> What concerns me at the moment is your treatment of [the] equation in which I say that the demand for money is identically equal to zero. (Incidentally, you repeatedly refer to this as an *assumption* of my system. This is absolutely not true. As you can see quite clearly . . . it is an implication of the assumptions of [the system].) . . . The statement that the demand for money is equal to zero identically in the prices is a meaningful statement. Economists expect economic variables to be dependent upon prices. This statement tells us that in [the classical monetary system] the demand for money has this property, that it is always zero regardless of what the prices may be. For an economist, this is an important piece of information. (emphasis in original)

Over a year later, in early September 1951, Phipps and Patinkin met in Chicago to discuss their disagreement. However, there was no resolution. On 18 September 1951 Phipps picked up the correspondence once again.

> You seem to believe, and lead the reader to suppose, that your system is the only one possible on mathematical grounds in which the utility functions do not contain money. However, there are others; . . . for

example, there is a system in which [price of money equals zero], a system which you violently reject.

Thus, you have placed an "iron curtain" around your solution. The reader is never told that there are many others which satisfy the broad conditions of the problem. And finally, you have arrived at your conclusions by calling things by their wrong names and giving the wrong reasons for reaching these conclusions.

From his own perspective, Phipps of course was correct, though "wrong names" is childishly essentialist and the "iron curtain" allusion is on the edge of good taste. The system with which Patinkin was concerned had a context, a history, and an established place in the discursive practices of monetary theorists. It was not in fact an arbitrary system, but rather one in which certain distinctions could be made (numeraire money, money of account, fiat money, money as an asset, etc.). That Phipps saw correctly that Patinkin's system was not unique though was of little consequence. Alternative assumptions, which seemed to satisfy logical coherence requests, were in fact economically incoherent, meaning that economists simply were not concerned with them. The argument Phipps constructed was similar in kind to demand analysis put forward in Griffith Conrad Evans's *Mathematical Introduction to Economics* (1930), which argued that utility theory was meaningless, and which simply postulated various closed form demand functions at the market level. Economists came to ignore such a discussion, disconnected as it was from their own analytic, and discursive, practices (Weintraub 1998).

The correspondence on this issue ended on January 5, 1952 when Patinkin wrote: "I do not think that there is any point in writing a detailed criticism of your letter. In it you continue to insist on misinterpreting arguments which to me seem to be quite clear. I was not able to make these points clear during the long evening which we spent together in Chicago. I doubt very much that I shall be able to do by means of correspondence."

Neither with a whimper, nor a bang, but rather with a "kiss off, Cecil," did thus Patinkin exit. The issues indeed never were joined, perhaps because neither could adopt the other's perspective. One of the exercises psychologists use to break a communication impasse between two individuals is to get each to argue the other's position, to get them to switch perspectives, as it were. Seeing an issue as another sees it is empathetic understanding, and such empathy sometimes engenders enhanced amenability to changing one's mind. If, however, neither can see the opponent's argument, cannot state it fairly, cannot appreciate the contingent circumstances from which it arises and takes on both coherence and merit, communication is doomed.

Though in his final letter Patinkin finally dismisses Phipps, it is striking that Patinkin allowed the correspondence to continue as long as he did.[10] Such patience may be attributable to Patinkin's early schooling at a yeshiva. As he writes in the postscript of "The Training of an Economist," the years in a yeshiva instilled in him "the patience to spend hours reading and rereading a difficult text in the Talmud, paying attention to its context, and to the nuances and minor differences of phrasing that provide clues to its meaning" (388). This experience further instilled in him "the confidence that if I were patiently to read and reread a text in economics in this way, I would eventually succeed in penetrating to its meaning, in understanding the intent of its author" (388). This leads one to ask whether Patinkin ever did understand Phipps's intent. It is possible that Phipps's criticisms did have some residual effects on Patinkin's future work. In the preface of his *Money, Interest, and Prices*, Patinkin comments on the debate that took place in the journals regarding his first two articles. "Needless to say, the necessity for preparing these replies was an invaluable stimulus toward a constant reworking and improving of the argument and its exposition" (viii). He does not, however, mention Phipps by name. Perhaps recalling Phipps's criticism of the numeraire, Patinkin later in the book differentiates between money as an abstract unit of account and fiat paper money. The former "serves only for purposes of computation and record keeping. This unit has no physical existence; that is, it does not coincide with any of the goods which exist in the economy" (18). Patinkin then uses the same example that he expressed to Phipps: "Perhaps the most familiar [example of abstract money] is the guinea in present-day England" (19). If this was an effort to addresses Phipps's criticisms, the effort was lost on Phipps. In response to the publication of *Money, Interest, and Prices*, Phipps took the time to contact the publisher of the book to voice his displeasure with the quality of Patinkin's work.

The belief that mathematics is a language that can be translated into economics held sway in the middle of the twentieth century. The correspondence between the economist Don Patinkin and the mathematician Cecil Phipps provides one case in which the relationship between mathematics and economics is more complex than is suggested by the midcentury mathematical economists.[11] Economists and mathematicians are trained in different discursive practices, each containing unique (though perhaps overlapping) persuasive techniques and rules of assessing evidence. The correspondence between Phipps and Patinkin demonstrates that these differences at times make it difficult, if not impossible, to resolve disagreements on the validity and application of a mathematical proof in economics. The difficulties experienced by our protagonists arise because the different communities of mathematicians and economists do not share histories, training, techniques of persuasion, rules for assessing evidence, or even languages.

As Stanley Fish presents it, "Interpretive communities are made up of those who share strategies not for reading but for writing texts, for constituting their properties. . . . [Moreover], since the thoughts that an individual can think and the mental operations he can perform have their source in some or other interpretive community, he is as much a product of that community (acting as an extension of it) as the meanings it enables him to produce" (Fish 1980, 14). Understanding that "texts" here are to be understood as the material products of these communities, their theorems, descriptions, analyses, arguments, etc., we can appreciate how economists' understanding of the role of mathematics is associated with their socialization as economists. That is, the discursive practices which emerge from the training economists receive are different from those of mathematicians, and this itself may make resolving disagreements between economists and mathematicians on the validity and application of a mathematical proof difficult if not impossible. Their conceptual frameworks are effectively incommensurable.

Phipps, who was not an economist, could not see or understand the context in which one assumption was acceptable in economics and another was not. Patinkin, who was not a mathematician, could not appreciate the content-less approach to ideas of consistency and inconsistency, or the belief that one had freedom to choose among alternative axiomatizations in order to produce mathematically interesting results. Despite their mutual talk of "translation," neither spoke the other's language. More than just conceptual incommensurability was involved. With apologies to

Strother Martin's character of the warden in *Cool Hand Luke*, "What we have here is a failure to communicate": the failure was necessary since the incommensurability was multilayered and inevitable. Socialized differently, educated differently, with different knowledge of monetary theory and mathematics, and with different objectives in argument, each worked with a different boundary between economics and mathematics, and so each conceived of mathematical economics differently. For Patinkin, mathematics could be used in an autonomous economics to formalize economic arguments, and to facilitate the making of rigorous deductions. For Phipps, mathematics was the doppelgänger for all discursive practices that presumed to argue from assumptions to conclusions: any such argument could be translated into mathematical terms—provided with an interpretation in mathematics—and be made rigorous thereby, or be shown to be incoherent. Phipps's view prefigured the position attributed decades later to the Bourbakist Renè Thom who claimed (defending catastrophe theory) that "in the future, only mathematicians will have the right to be intelligent."[12]

6 Equilibrium Proofmaking

with Ted Gayer

There is no Algebraist nor Mathematician so expert in his science, as to place entire confidence in any truth immediately on his discovery of it, or regard it as anything but a mere probability. Every time he runs over his proofs, his confidence encreases; but still more by the approbation of his friends; and is rais'd to its utmost perfection by the universal assent and applauses of the learned world.
—David Hume, *Treatise on Human Nature*

Each year, new economics Ph.D. students learn the proof of the existence of a competitive equilibrium as if a rite of passage. From the utility maximizing behavior of consumers and the profit-maximizing behavior of firms, neophyte economists soon can demonstrate that under certain conditions there exists a competitive market-clearing general equilibrium price vector. While there are a number of proofs that establish the existence of such an equilibrium, the validity of these proofs is indubitable. Indeed, economists with even scant knowledge of the history of economics can identify Kenneth Arrow and Gerard Debreu's 1954 *Econometrica* paper as having provided the proof that settled the issue.

That paper, "On the Existence of an Equilibrium for a Competitive Economy," appeared to bring closure to an argument that was (at least) two centuries old. The paper was cited in the award of the Nobel prizes to both Arrow and Debreu. The canonical account of the context and origin of the Arrow-Debreu paper suggests that its history may be traced through a series of different lines in several literatures, in several disciplines, on at least two continents, all of which converged to publication in 1954.[1]

The Arrow-Debreu model was a major accomplishment; it presented an economy composed of individual, self-interested agents—both utility-maximizing households and profit-maximizing firms—pursuing their own self-interest and whose actions produced an equilibrium in which all choices were potentially reconciled. Put briefly, the pursuit of individual self-interest could lead not to social chaos but to a coordinated social order. But how did a piece of work in mathematical economics actually settle an economic question? How did it come to pass that a particular paper, in a journal at that time read by very few economists, came to be accepted as having established a foundational truth about market economics? These are not questions economists typically ask. "The theorem proves that..." is enough information to persuade economists that the knowledge associated with the theorem is secure knowledge. Professional economists are confident about the result and the implications of the equilibrium proof, and no one needs to attend to the means of its construction: the validity of the equilibrium proof is incontrovertible. Economists-in-training must learn that the existence of a competitive equilibrium has been proved. All economists can make use of the proof of that result without subjecting it to incessant challenge and reassessment.

Scientists must take some components of their research as given; intellectual paralysis awaits the scientist who seeks to reopen every foundational issue every day.[2] For most economists the competitive equilibrium proof is a tool to use with little regard to how the tool was constructed. Those who study science use the idea of a "black box" for settled results that are locked up and impenetrable, and thus closed to current investigation.[3] For every science, black boxes are both healthy and necessary. But how do novel ideas get closed up into black boxes? By what means does a new claim of knowledge gain acceptance within a scientific community? Today, decades after the publication of the Arrow-Debreu proof, it is relatively easy to view it as both immutable and uncontroversial. Yet how was its validity assessed initially? More generally, how does a scholarly community determine that a proof is valid, especially when the proof is highly complex and when there are few people in the community with the technical skill to understand the proof? And what might "understanding a proof" entail?

We will address such questions by examining in some detail the circumstances that surrounded the assessment and the publication of the Arrow-Debreu paper. To examine the circumstances in which the proof became common knowledge does not, of course, diminish the proof itself. Understanding the world in which Newton lived and made his contributions offers insight into the formation and acceptance of his contributions without denying the truth of his theories. Likewise, by considering the community in which the Arrow-Debreu paper appeared, we seek to understand how the economics community assessed and established a claim to knowledge, not to denigrate Arrow and Debreu's exceptional contribution. To borrow from Bruno Latour (1987, 4), we will explore the distinction between "ready made science," which is represented by the valid and true equilibrium proof, and "science in the making," which is represented by the proof's construction, assessment, and initial acceptance by some members of the community. Thus this chapter will seek to uncover both the context in which, and the process by which, Arrow and Debreu's proof moved from being a novel claim within the small community of mathematical economists to being an established truth among the much larger community of economists.

Closing the Black Box

When was the black box closed? Did the economics community universally accept the validity of the proof immediately upon publication, or did it take several years for the proof to become known? One way to gauge its acceptance within the profession is to examine the standard microeconomics textbooks used in the training of advanced undergraduate students, and new Ph.D. students.[4]

If we start in the 1940s, years before the publication of Arrow and Debreu's proof, we find Sidney Weintraub's take on the existence of a competitive equilibrium. In his 1949 book, *Price Theory,* he wrote:

> Presumably we might be ready to concede that particular equilibrium analysis divulges some fundamental tendencies in the economy, the end-results of market processes that secure a balance. But we are much more reluctant to concur in the view that all markets are in balance

simultaneously; admitting the tendency in individual markets is still a long way from subscribing to the proposition for all markets simultaneously, over any period or even any moment of time. But once we acknowledge that in each particular market and in each sector of the economy that there are certain equilibrating forces at work, there is no sensible reason to shrink from the view that the entire system, or a good portion of it, can settle down in an equilibrium of supply and demand. . . . Nevertheless, whatever violence the idea of general equilibrium does to our sense of reality, and even if we entirely reject it as an artificial image of the economic world, it is still incumbent upon us to demonstrate the conditions that need to be satisfied for the general equilibrium of production and consumption, and to explore the interdependence among markets. (Weintraub 1949, 127)

Thus, while not referring to a "proof" of the existence of an equilibrium, Weintraub informed the young Ph.D. student that "there is no sensible reason to shrink" from the view that an equilibrium exists.[5] And while questioning the "sense of reality" of the general equilibrium model, Weintraub stressed the importance of demonstrating the conditions that satisfy the existence of an equilibrium. This demonstration occurs a few pages later:

Rather than rely on the verbal proof that general equilibrium is conceptually possible, the mathematical proof rests on the demonstration that for each price that is to be determined we have an equation. *If the number of equations is equal to the number of unknowns then the results are deemed to be determinate*; the counting of equations gives evidence that there is a set of prices that can establish simultaneous equilibrium in the several markets. Other properties of the structure of equations, such as the demonstration that the equations permit of only a unique set of prices, are regarded as a problem mainly of mathematics rather than of economics. The economic interpretation is often fairly simple. (130; emphasis added)

In 1949 the proof of the existence of an equilibrium (as presented to Ph.D. students) rested on the equivalence of the number of equations and the number of unknowns in the system of equations. This line of argument had a long history in economics, going back to Walras, and was reiterated in textbooks like those of Bowley and treatises like Hicks's *Value and Capital*. Even though a small number of economists in the 1930s had understood that establishing the existence of an equilibrium was a difficult mathematical problem, and even though there were some notices of the work by Wald and von Neumann to appear in the larger literatures, such analyses seemed not to have "crossed over" as it were into mainstream economics, instead being relegated to the backroom of "mathematical economics."[6] As our concern is with the transition process by which the understanding of a small coterie became the knowledge of the larger community, we will want to see how the idea of counting equations and unknowns was discarded and replaced with the existence proof presented by Arrow and Debreu.

Another competing textbook used at that time was George Stigler's *The Theory of Price*. In the first edition of 1946, Stigler mentioned general equilibrium briefly in a subsection of the introductory chapter. He first voiced skepticism concerning general equilibrium studies,[7] stating that "*general* equilibrium is a misnomer: no economic analysis has ever been general in the sense that it considered *all* relevant data. . . . The most that can be said is that general equilibrium studies are *more* inclusive than partial equilibrium studies, never that they are complete" (28; emphasis in original). He had little to say concerning the existence of an equilibrium, writing that "the outstanding characteristic of the conditions of equilibrium is that they are equal in number to the unknown quantities and prices which are to be determined. The conditions are, in mathematical terminology, the equations of the economic system, and prices and quantities are unknowns" (30).

In a substantially revised edition of 1952, Stigler added a chapter (the last one) on "General Equilibrium." While it was still two years before publication of Arrow and Debreu's proof, Stigler mentioned that "some beginnings have been made to a theory of general equilibrium" (287). This suggests that he had some idea that proof by counting equations and unknowns was unacceptable to some theorists.[8] In the meantime, as was done in Weintraub's text, Stigler set up the demand functions and supply functions, then counted the number of equations and the number of unknowns in the system, and concluded:

> The set of prices and quantities satisfying the equations constitute a general equilibrium: we have simultaneously fulfilled the conditions that quantity demanded equals quantity supplied in every market, taking full account of the fact that supply and demand in each market depend (in ways fixed by consumer and producer behavior) upon all the prices in the economy. A change in the demand for any commodity, or in the quantity of a productive service, or in any production coefficient, or the fixing of one price by fiat, will affect all other prices and quantities. (294).[9]

Though ignored in Stigler's textbooks, Arrow and Debreu's *Econometrica* article was cited in the 1958 first edition of *Microeconomic Theory: A Mathematical Approach*, by James M. Henderson and Richard E. Quandt. The chapter on "Multimarket Equilibrium" contained a subsection on "Existence Theorems," which summarized, with few details, the new proof. It is here that we first see the claim in a textbook that "Arrow and Debreu have proved that a competitive equilibrium solution exists" (155). Thus, doctoral students at the time were taught that an equilibrium exists under certain conditions, but they were not taught the proof itself in the textbook:

> Arrow and Debreu have considered the problem of existence for abstract multimarket systems similar to the one presented [previously]. . . . They employ set-theoretical techniques rather than differential calculus. Their assumptions for the first of the two cases which they consider are approximately as follows: (1) no firm realized increasing returns to scale, (2) at least one primary factor is necessary for the production of each commodity, (3) the quantity of a primary factor supplied by a consumer cannot exceed his initial endowment, (4) each consumer's ordinal utility function is continuous, (5) consumers' wants cannot be saturated, (6) indifference surfaces are convex with respect to the origin, and (7) each consumer is capable of supplying all primary factors. Arrow and Debreu have proved that competitive equilibrium solutions exist for all systems that satisfy these assumptions. They weaken assumption (7) in the second of the existence proofs. (155)

In the preface to their 1971 second edition, Henderson and Quandt listed the "proof for the existence of equilibrium in a competitive economy" among the "new material that appeared in the economic literature since the publication of the first edition or was considered too new or difficult for inclusion at the earlier time" (v). In this second edition the authors added a subsection on "The Existence of Equilibrium" to the chapter on "Multimarket Equilibrium." Unlike the first edition, the authors went on to provide a detailed account of a proof of the existence of an equilibrium for "particular sets of excess demand functions" (178) and for "the general problem of existence for a short-run version of the production and exchange system presented in [a previous section]" (178). Instead of using Arrow and Debreu's proof, they focused on Brouwer's fixed-point theorem to prove the existence for the restrictive case. For the general case they only offered an outline of Debreu's use of the Kakutani fixed-point theorem from his (1959) *Theory of Value*.

Thus, as early as 1958 Henderson and Quandt were instructing economics students that under certain assumptions, Arrow and Debreu "have proved that competitive equilibrium solutions exist." Within a few years of the publication of Arrow and Debreu's *Econometrica* paper, the validity of their proof had gained widespread acceptance within the community of economists, although the details were not presented to the students in microeconomics textbooks. And by 1971, not only was the new proof of existence of an equilibrium universally accepted within the profession, but students were presented both with the details of a special proof based on Brouwer's fixed-point theorem under certain restrictive demand assumptions, and also with an outline of Debreu's less restrictive proof.[10]

If we are correct in assuming that Ph.D. textbooks reflect the consensus of what constitutes knowledge within a discipline (or instantiates the paradigm of normal science), then we can infer that the Arrow-Debreu proof was generally accepted as having established the existence of a competitive equilibrium by 1958, which was but a few years after publication of the article. If one further believes that it was the publication of the proof in *Econometrica* that signaled the acceptance of its validity by the community of mathematical economists, and thus convinced the broader community of the truth of the sentence, "There exists a competitive equilibrium market clearing price vector," then the historians' task is to examine the process by which *Econometrica* assessed the proof.

We do not mean to suggest that the broad acceptance of the validity of the proof was a discrete event, occurring one day at a seminar presentation or sometime in July of 1954 when the *Econometrica* volume containing the article was circulated. Indeed, we believe that the acceptance of a novel claim of knowledge is a dynamic process, as the lid to the black box descends, gains momentum, and ultimately slams shut. In 1949, the consensus within the nonmathematical economics community was that a competitive equilibrium existed, and it could be established by counting supply-demand equations and price unknowns. By 1958 Henderson and Quandt confidently asserted that Arrow and Debreu had proved the existence of a competitive equilibrium, implying that the result had not been established as true earlier.

Writing and Submitting the Paper

The story of the Arrow-Debreu paper is relatively well known (Weintraub 1983; 1985). The history of "existence of equilibrium" is a story of Abraham Wald's work in Vienna, following Schlesinger and perhaps Remak (in Germany), John von Neumann's 1936 masterpiece, and separate lines of attack that developed with Kenneth Arrow at Hotelling's Columbia, Gerard Debreu from Bourbaki's loins at the École Normale Supérieure, and Lionel McKenzie's "retooling" rebirth at Tjalling Koopmans's and Jacob Marschak's Cowles Commission over twelve months in 1949–50 (Weintraub 1985, 98–100). The Arrow-Debreu collaboration emerged from the work each of them did separately, and although their times at Cowles did not overlap, they eventually learned of each other's activity through the organization's working policies: in 1950–51, Debreu was given the paper by Arrow on "the fundamental theorem of welfare economics" to referee for the internal Cowles publications system, and was asked to "comment on the substance of the paper" (Weintraub 1983, 28). That paper was in fact quite similar to Debreu's own paper, written prior to June 1950, which was to appear in the July 1951 *Econometrica* as "The Coefficient of Resource Utilization" (ibid.). Both the Debreu and the Arrow papers set up the structure of the competitive model in a form that was to be used by each, in their own next papers, to establish an equilibrium. That is, by early 1951 both Arrow and Debreu were working with a model of an economy in which the definition of a competitive economy was developed in a fashion consistent with an approach for examining the equilibrium price system for that economy.

It was then for Kenneth Arrow that:

> According to my recollection, someone at RAND prepared an English translation of the [Wald] *Ergebnisse* papers to be used by Samuelson and Solow in their projected book (sponsored by RAND), which emerged years later in collaboration with Dorfman. I read the translations and somehow derived the conviction that Wald was giving a disguised fixed-point argument (this was after seeing Nash's papers). In the fall of 1951 I thought about this combination of ideas and quickly saw the competitive equilibrium could be described as the equilibrium point of a suitably defined game by adding some artificial players who chose prices and others who chose marginal utilities of the income for the individuals. The Koopmans paper then played an essential role in showing that convexity and compactness conditions could be assumed with no loss of generality, so that the Nash theorem could be applied.
>
> Some correspondence revealed that Debreu in Chicago [at Cowles] . . . was working on very similar lines, though he introduced generalized games (in which the strategy domain of one player is affected by the strategies chosen by other players). We then combined forces and produced our joint paper. (as quoted in Weintraub 1985, 104)

In similar vein, Gerard Debreu recalled that:

> [It] was when [the Koopmans monograph] was published that I learned of the existence of A. Wald's papers on general economic equilibrium, and only when the English translation appeared in *Econometrica* [October 1951], did I get acquainted with its contents. At that time, in the Fall of the 1951, I was already at work on the problem of existence of general economic equilibrium. . . . The influences to which I responded in 1951 were the tradition of the Lausanne school and, in particular, the writings of Divisia, Hicks, and Allais; the theory of the games and, in particular, the article of J. Nash; the [paper on fixed points by] Kakutani and the [1937] article of von Neumann . . . [as well as] the linear

economic models of the Cowles Commission monograph. (as quoted in Weintraub 1985, 103–4)

We also know that Lionel McKenzie, who was teaching at Duke University in the late 1940s, went to Chicago initially to take a summer course at the Cowles Commission in 1949. At that time, McKenzie became aware of the programming material that Koopmans was developing for the Cowles conference, and he then began work on the paper on Graham's model of world trade that became his contribution to the existence of competitive equilibrium literature. McKenzie remembered that

> My paper and the paper of Arrow-Debreu, which were developed completely independently, were presented to the 1952 Chicago meetings of the Econometric Society. I recall that Koopmans, Debreu, Beckmann, and Chipman were at my session. The Arrow-Debreu paper had been given the day before and I had stayed away. However, Debreu rose in the discussion to suggest that their paper implied my results. I replied that no doubt my paper also implied their results. As it happens, we were both wrong. Debreu [has told me] he spoke up after asking Koopmans's advice before the session. Later in his office, Debreu gave me a private exposition of their results. (as quoted in Weintraub 1985, 103)

We need to be clear about this chronology, since we are interested in the dissemination of Arrow and Debreu's proof before they submitted the article to *Econometrica*. On 27 December 1952 the Arrow-Debreu paper, and the McKenzie paper, were presented in an open session, and heard at least by Koopmans, Beckmann, and Chipman. We believe that several members of the Cowles staff knew of and perhaps had read the papers, at least the Arrow-Debreu paper, and the results it contained, so that Marschak and others were aware of the paper, even if they had not read it.[11]

Since the paper was submitted for publication under the dual cover page note "Technical Report No. 8 prepared under contractN6onr-25133 (NR-047-004) for Office of Naval Research, Department of Economics, Stanford University, Stanford California and Cowles Commission Discussion Paper in Economics No. 2082 prepared under contract nonr [sic] - 358(01) (NR-047-006) for Office of Naval Research" it is clear that before it was submitted for publication to *Econometrica*, probably in the first week of June 1953, it had to have been read by internal Cowles referees as well as those monitoring/refereeing the Office of Naval Research contract products. We have then a quite usual scholarly time frame for that pre-fax, pre-Xerox period: the paper was presented in late December 1952, it must have been retyped, and read by people at Cowles and perhaps Stanford in the winter and spring of 1953, and sent back and forth between California and Illinois with changes and corrections and emendations prior to submission to *Econometrica* on or around 1 June 1953.

Our evidence for the submission date is a letter, dated 15 June 1953, from Robert Strotz (managing editor of *Econometrica*) to Nicholas Georgescu-Roegen that dealt with three separate matters.[12] The third paragraph reads:

> I am enclosing three copies of a manuscript submitted by Arrow and Debreu which falls in your department [as Associate Editor]. I hope you will be good enough to arrange for the refereeing of this paper and to advise me on it. I should mention that a rather similar paper was submitted some time earlier by Lionel McKenzie and that it has not yet completed it [sic] processing. As a matter of fact it is being read at present by Leo Hurwicz and John Nash. I suppose, therefore, that these two readers should not be burdened further with the Arrow-Debreu paper.

Thus, Georgescu-Roegen was given little advice on whom to choose as referees, only being told not to choose Hurwicz or Nash (and by implication, not McKenzie). The choice of referee is complicated by the tradeoff between finding a qualified referee and finding an impartial referee. It is a rare referee who reads every line and every calculation of a paper. As noted by the mathematicians Philip Davis and Reuben Hersh, "[Only one] whose interest and training are very close to the author's would be willing and able to do this kind of checking" (Davis and Hersh, 61). Yet a referee with such interests may be prejudiced toward publication and thus might be a poor referee.

Reading that past from this present, the process at *Econometrica* was troublesome. If the associate editor had been charged to find individuals with little or no connection to either Arrow or Debreu to referee the paper, he was going to find that to be a difficult assignment. Certainly all of the Cowles people were "disqualified." Likewise the people at RAND who were connected to Arrow by that time were not going to be able to help. But there were not many mathematical economists in 1952 outside those groups. Except for a very few places like Chicago, MIT, or Stanford, the community of mathematical economists hardly existed in the early 1950s. For example, we know that Sidney Weintraub, who had taken at most one calculus course in his entire undergraduate and graduate career, was implored by his chairman Raymond Bowman, and agreed, to teach the graduate course in mathematical economics at the University of Pennsylvania in 1950–52. Finding a mathematical economist to appraise the mathematically complex paper by Arrow and Debreu was not easy since mathematical economists tended, like individuals in any other marginalized subdiscipline, to send their writings to each other before submitting them for publication. Their papers were most often presented in conferences sponsored by the Cowles Commission, the Econometric Society, or by the RAND Corporation.[13] With this in mind, Georgescu-Roegen's choices for referees appear less curious.

The Referees

The two referees selected by Georgescu-Roegen were William Baumol of Princeton University's Economics Department, and Cecil Glenn Phipps of the University of Florida's Mathematics Department.[14] The request to serve as a reader went out on 23 June, and on 17 July 1953 Baumol duly sent his report off to Georgescu-Roegen. After some preliminary comments on another matter, Baumol wrote:

> I think this is a very important paper indeed, and have not the slightest doubt that it ought to be published. My only major suggestion is that, despite its length [forty-seven double-spaced pages], it would be useful to the reader to have something more explicitly said about the fundamental lemma on page 16.[15] So much is built on it and the reference to Debreu's derivation [in his earlier 1951 *Econometrica* paper] is not readily accessible. The extra space which would be required would be well worth it.

Baumol then went on to note four specific "minor suggestions" on issues like missing bracket signs and omitted circumflexes. We may thus assume that Baumol read the paper carefully enough to do some proofreading, and he believed that the paper was "very important" although he did not, in the report, discuss why this might be the case.

The report of the second referee, Cecil Phipps, has not been found. We do, however, have the account of that report that Georgescu-Roegen provided to his editor, Robert Strotz, in a letter dated 8 October 1953: "Phipps has complained many times that the mathematics of economists is faulty and I thought he would thoroughly check the mathematics of the argument. He did not do as I had hoped. Instead, he concentrated on the discussions of the axioms. Phipps is emphatically against publication, until the paper is revised. I think his comments should be sent to the authors. Perhaps they will be able to make more of them than I was."

Given the trouble he was causing, it is not unreasonable to ask "who was Phipps, and why had he been selected?"[16] We have met Phipps in the previous chapter so we will not repeat that treatment here. We reiterate though that there is no evidence that Phipps published any mathematical research. Rather he was a mathematics teacher in a small, teaching-oriented, segregated Southern public university. Somehow, though, following military service in the Second World War, Phipps got interested in mathematical economics, and became the leader of a small group of faculty and graduate students with common interests at the University of Florida. We have no written record of the meetings of this group, but we have the evidence of Phipps, who refers to his being "a member of a group" in his correspondence with Don Patinkin discussed earlier.

His student Miller's dissertation, "The Mathematics of Production and Consumption in a Static Economy," is an excellent window into Phipps's views on general equilibrium theory.[17] Miller writes of the "errors and misconceptions" that occur in the new science of mathematical analysis as applied to economics, and tells us that "nowhere in the literature have I been able to find a complete and correct mathematical treatment of the general case of production and consumption" (ii). Miller then proceeds to develop a theory of consumption, and production, and to link them with a theory of competition to produce a solution of what are, in effect, the equations of general equilibrium. The problem though is that Miller's analysis, written at the end of the 1940s for the 1951 thesis, is incoherent with respect to then current economics. And since Miller was supervised closely by Phipps, and since it was Phipps who was to take it upon himself to stop publication of the Arrow-Debreu paper based on his own understanding of how to do general equilibrium theory, it is worth pausing another moment in our story to reconstruct Phipps's beliefs, which it is fair to assume are expressed by Miller, about this cornerstone of mathematical economics.

The thesis had seven chapters. The first three contained routine reviews of optima and restricted optima, and homogeneous equations. Also, there is a discussion of what are termed "independent functions," where a "set of functions is dependent if, when values are assigned to some of the functions in the set, the values of one or more of the other functions are determined." The distinction between local and global properties seems to be ignored here in this imitation of linear independence. In any event, following this basic material, Miller goes on to apply it to economics, nevertheless ignoring the entire published literature in mathematical economics: although he has references at the end to Samuelson, R. D. G. Allen, and Hicks, etc., Miller seems not to use these books in any of his chapters. For example, he writes that marginal productivity theory is not a theory that is "both complete and correct from a mathematical point of view" (58), yet he does not point to any mistake in any other author and develops what he calls his theory (which in fact was quite standard in economics textbooks like Weintraub's and Stigler's) with only two variables! Moreover, for a mathematical treatment there was no recognition of the problems associated with non-negativity constraints, and this after the Cowles conference on programming.

The thesis builds to a final chapter in which the material on production, and that of consumption, is joined to produce a model of a closed competitive economy. Miller sets out to establish "the equilibrium point at which the economy has 'settled down': i.e., of determining the amounts of the n X's [quantities] and the n-1 price ratios of these commodities in terms of the fixed capital assets and their distribution. The solution will also embody known production and utility functions" (142). What follows is a careful rendering of all the equations of utility maximization, and profit maximization, together with assumptions about competitive markets, albeit with no recognition of the problems associated with non-negativity constraints on prices. Miller ends up with an enormous number of equations, and one less price. He then states that he can eliminate variables ending up with 2n-1 equations that suffice to determine the n amounts and n-1 price ratios, and thus establish the competitive equilibrium mathematically.

As with the textbooks by Weintraub and Stigler in the 1940s, Miller's and Phipps's view of a proof of a competitive equilibrium rests on counting equations and unknowns. Their idea of what constituted a proof is similar to the closed black box of the previous decade. Given that Arrow and Debreu's proof changed this conception of the black box of a standard proof, we can anticipate Phipps's reaction (which we will discuss later) to this challenge of what he perceived as irrefutable. It takes one's breath away. In 1951, in a mathematics department, in a thesis with references to Samuelson, Cecil Phipps and his student William Miller have recreated the equation counting argument used in microeconomic textbooks in the 1940s and sneered at in the open literature by Morgenstern ten years earlier (Morgenstern 1941). It is as though Wald had never solved the problem stated earlier by Schlesinger, and that von Neumann's paper had never been published, let alone translated into English. This was to be Phipps's contribution to the existence of general equilibrium literature, a failure to read the literature.[18] As Patinkin was to write to him, at about that same time (12 April 1950) in another context: "I am firmly convinced that you and your group must spend at least one or two years learning the basic fundamentals of mathematical economics before any worthwhile criticism will be forthcoming."

In any event, this was the intellectual framework that was to shape referee Phipps's response to the Arrow-Debreu paper.

The Decision to Accept the Paper

The first stage of the review process thus ended with Georgescu-Roegen's report to Strotz of 8 October. That report shows that the associate editor did his own appraisal of the paper, effectively refereeing it himself in light of the two reviews he had received.[19] He is quite certain about his judgment, and his six-page single-spaced letter deserves to be quoted at length:

There is no doubt in my mind that the paper deserves to be published. Therefore the comments which follow should be interpreted simply as suggestions . . . and not as belittling the authors' contribution.

After I received the manuscript, I read it superficially to decide to whom it should be sent for refereeing. My first impression was that the mathematics was rather intricate even for the top econometricians, and this opinion was reinforced after having recently read the article more carefully. In addition, the mathematics and the economics are so much inter-woven in the argument that I found it difficult to think of many referees who would be at the same time economists and mathematicians so that the critical reading of the paper would not impose upon them a tremendous task. I have asked Baumol and Phipps to comment upon it.

Georgescu-Roegen goes on to present his views on Phipps's report, as noted above, and then states Baumol's comments in favor of publication. He says he is "glad to have one of the referee's opinions to add support to my favoring the publication, so much more since this comes from an econometrician like Baumol." Nonetheless, he informs Strotz that Baumol's remarks are "trivial," and that "he did not check the argument in detail." He continues, "I do not blame him for choosing not to spend the rather considerable time required by the job." He also admits to Strotz that he, too, did not give the manuscript an exceedingly careful reading, but instead based his decision at least in part on the reputations of the authors. "I also decided that to go over the manuscript as I used to do in the past would have taken too much time. I felt that the following remarks would be more valuable to the authors than a thorough checking of the mathematics by me. I have the highest opinion of the authors and I trust Debreu's mathematics, yet I recommend that somebody check the mathematics. This could be done while the authors revise the present version, thus saving considerable time." It is not clear whom Georgescu-Roegen expected to "check the mathematics" of this admittedly complex paper.

Before going on to present Strotz, who would be the one to communicate with the authors, with specific recommendations, Georgescu-Roegen would make the following plea for simplifying the paper: "Would it not be possible either to make the proof more elementary and simpler or to present it as elaborated consequences of other well-known theorems? I heard at Kingston[20] the paper given by McKenzie and was impressed by the very small place occupied by the technical mathematical proof in the argument."[21]

In his next set of seventeen numbered remarks to Strotz about the paper, covering four pages, Georgescu-Roegen more or less set the stage for many of the issues which would be subsequently involved in methodological discussions of what has come to be called the Arrow-Debreu model.[22] He notes for example (point #3) that the authors call one of their assumptions "highly unrealistic" and suggests that it be shortened and given less emphasis. In point #1, he is clear in his view that the paper should separate the "mathematical proofs of the abstract lemmas and theorems from the economic interpretation of the result," a call to rethink, as it were, the nature of an argument in mathematical economics.

Other points question the relation of the model to Leontief's model, or the issue of stocks versus flows, or the issue of the number of firms being fixed in advance of the equilibrium discussion. In point #8, for instance, he notes that the paper explicitly avoids the question of uniqueness of equilibrium, and suggests that a similar mention be made about the stability of equilibrium. Of most interest to future methodologists perhaps is #6:

> The paper leaves the reader with the definite impression that the existence of equilibrium for an economic system requires rather strong assumptions. If one would like to derive some realistic conclusion from this, this conclusion would be that very likely the real system would be deprived of such assumptions and of an equilibrium, also. What is the reaction of the authors to such an interpretation?

The associate editor's report was duly sent on to Strotz. We do not have any follow-up letters to the authors, but can surmise that they were given the gist of the reports. We can also surmise that Strotz's conditions for final acceptance and publication were based on Georgescu-Roegen's letter. A comparison of the draft version in Georgescu-Roegen's files and the final published article shows that there were virtually no changes to the article between submission and publication. We believe that the response to the authors and the resubmission was done over the course of the next several months, and that the final version of the paper was ready by spring 1954, and that the editor so informed the referees. That timing is then consistent with the remarkable letter, and enclosure, that Strotz received at the end of the summer of 1954.

An Objection to Publication and *Econometrica*'s Response

On 18 September 1954 Cecil Phipps submitted a letter to the editor of *Econometrica*, Robert Strotz, criticizing the validity of Arrow and Debreu's article. In his cover letter Phipps expressed his displeasure with *Econometrica* for having publishing the article. "I do not feel that this article should go unchallenged before the readers of *Econometrica*. Otherwise, economists will accept its conclusions at face value and quote it in substantiation of other arguments, perhaps ones of economic policy affecting all of us." Phipps's letter makes little mention of the actual proofs used by Arrow and Debreu.[23] Instead, he criticizes the way they set up the model of a competitive economy, their definition of an equilibrium, and some of their assumptions about consumers and firms. Phipps begins by claiming that there "are only three parts to the problem instead of the four into which the authors divide it. The first concerns the individual firm whose inputs and outputs are functions of the fixed set of prices as parameters. The second concerns the individual consumer whose income is determined by the labor he performs and the material he has or received. . . . The third part . . . may be stated as follows: If the differences between the demand and production of all but one of the commodities are specified, can the prices at which these differences exist be found from the excess supply functions?" Thus Phipps offers what he believes to be the proper way of establishing the existence of a competitive equilibrium; however, he does not offer a proof of his own.

Phipps also criticizes certain assumptions used by Arrow and Debreu. For example, he thinks it is incorrect to postulate that firms (consumers) maximize profits (utility) for a given set of prices. Instead, "the maximum in this case must be attained for any permissible set of prices, not just the final equilibrium prices as they state." He also claims that treating inputs as negative components "becomes very awkward when the output of one firm becomes the input of another. . . . The argument of the authors would have to be changed slightly to care for the change in signs." He disapproves of normalizing the vector of prices by requiring that the sum of its coordinates be 1, stating that it "has no connection with the question of a solution for these prices . . . [and] serves merely to give a unique value to the prices after the solution for the relative prices has been accomplished."

Before assessing the responses to Phipps's letter, it is useful to keep in mind Philip Davis and Reuben Hersh's observation about influences on referees' judgments: "Do the methods and result 'fit in,' seem reasonable, in the referee's general context or picture of the field? Is the author known to be established and reliable, or is the author an unknown, or worse still, someone known to be unoriginal or liable to error?" (Davis and Hersh 1987, 61). Two studies by Douglas P. Peters and Stephen J. Ceci (Peters and Ceci 1980; Ceci and Peters 1982) offer evidence that referees do consider some of the questions posed by Davis and Hersh. In their study of psychology journals, they found that a paper by an unfamiliar author at a low-status institution (two characteristics that Phipps fits well) is more likely to be rejected by the journal.[24]

There is also some anecdotal evidence, in economics, to support the claim that the prestige of the referee carries weight in the editorial decision. Paul Samuelson wrote of one occasion on which, after writing a critical referee report for the *American Economic Review*, the editor asked him if it was acceptable to give out his name to the author, for the author had stated that "I would like to know who the referee is. For if it is Milton Friedman, I must take it seriously." Samuelson replied, "I authorize you [the editor] to tell the author that the referee was not Milton Friedman" (Shepherd 1995, 20–21). Herbert Gintis recalled an instance when his paper received one five-and-a-half page, single-spaced, referee report suggesting publication, and a second referee report of thirteen lines that recommended rejection. Gintis claims that this second report was "vague, sloppy, and incorrect," yet the editor decided to reject since "the Board has great respect for the opinion of Referee #2" (ibid., 73). Thus, to the extent that editorial decisions are based on the prestige of authors and referees, we would expect Phipps to have a difficult time convincing *Econometrica* of his objections, independent of its merits.

In order to decide whether to publish Phipps's letter to the editor, Bob Strotz solicited the written opinions of Ragnar Frisch (editor of *Econo-*

metrica), Lionel McKenzie, Kenneth Arrow, Gerard Debreu, Hukukane Nikaido, Tjalling Koopmans, and Nicholas Georgescu-Roegen.

Lionel McKenzie was highly critical of Phipps's letter to the editor. On 28 September 1954 he wrote to Strotz, "This letter is extremely feeble and does not deserve serious consideration! . . . I think it would be a terrible thing to have this letter appear in *Econometrica*." His suggestion to Strotz on how to deal with Phipps's letter was "either (A) tell Phipps the material was not appropriate for a letter but you had it refereed as a note and it was rejected, [or] (B) tell him the material is inappropriate for a letter but if he wishes to submit a note you will then have the note refereed." McKenzie also claimed, as Georgescu-Roegen did earlier, that the complexity of the article precluded a careful examination on his part. He proclaimed, as had Georgescu-Roegen previously, an implicit trust in Debreu's mathematical abilities. "Even if there are correct points in it [Phipps's letter], they are no doubt trivial, and it would take me a month of Sundays to find them! Debreu, of course, is far too competent to commit such silly errors as Phipps seems to think he finds."

Ragnar Frisch's response to Phipps's letter was somewhat more sympathetic. While agreeing that "this letter contains much irrelevant and trivial talk," he tenuously suggested that Phipps might have a point. He wrote to Strotz on 28 September 1954, "I do not feel convinced that it is all sheer nonsense. I have a feeling that Phipps is perhaps touching upon some of the same fundamental difficulties that I have treated in my big paper to appear in *Économie Appliquée*. . . . I have a feeling that the kind of approach used by Arrow and Debreu could perhaps be criticized by an argument similar to the one I followed in this paper."[25] However, as Georgescu-Roegen and McKenzie did before him, he intimated that the complexity of the Arrow-Debreu paper would take too much time to examine the specific criticisms. "To find out whether this is actually so, is not a quick job, it would mean going through the paper in July 1954 *Econometrica* very carefully." Frisch was uncertain about how to advise Strotz on the matter, and wished to obtain "the reactions of Georgescu-Roegen, Lionel McKenzie, Tjalling Koopmans, and Gerard Debreu before reaching a final decision."

On 5 October 1954 Arrow and Debreu responded to Strotz about Phipps's criticism. They claimed that in order to prove his point Phipps must do one of two things: "1) to point out, with reference to page and line, where we make an inference which is not warranted by our assumptions or by logic, [or] 2) to present a model satisfying all our assumptions and having demonstrably no equilibrium in our sense." They dismissed Phipps by concluding, "As he does neither it is very difficult for us to take his comments seriously." While they believed Phipps's argument "exhibits throughout the grossest mis-understanding of our paper," they acknowledged the delicacy of the matter. "We understand that it is very delicate to suppress any scientific criticism and only ask for a chance to have a brief reply published alongside his letter if it is eventually accepted."

Nikaido similarly dismissed Phipps's letter. His 7 October 1954 letter to Strotz states that Phipps "has failed . . . to understand the version of Arrow-Debreu [sic] article; he seems to confuse argument of economic relevance with mathematical argument to confirm the former. It does not matter, in my opinion, whether mathematical arguments used to achieve the existence of economically relevant solutions admit some economic interpretation." Nikaido reiterated this sentiment later when he submitted comments to Strotz on Phipps's letter to the editor. In this 20 October submission Nikaido offered his point-by-point evidence that Phipps "has not succeeded in apprehending the version and the basic framework of the [Arrow-Debreu] article. In reading such an article as that of Arrow-Debreu, one should take much care that the economic formulation of a problem is not the same thing as the mathematical processing carried out to achieve a solution corresponding to the former."

On 19 October 1954 Tjalling Koopmans wrote up his opinion for Strotz. Koopmans acknowledged that "some of Phipps' comments point up inadequate explanations of the relation of the authors' (Arrow and Debreu) model to economic realty [sic] as well as yet unsolved problems." Nevertheless, he believed that Phipps "does not start from their premises to point out any specific errors in their chains of reasoning. Rather, he argues how he would have gone about this problem, and notes various differences, which he then describes as failures of the authors." Koopmans then gave a point-by-point analysis of Phipps's letter in an attempt to "help remove misunderstandings and thus conserve space in *Econometrica* for discussion of essential difficulties and unsolved problems."

By 3 November 1954 Strotz had received all solicited reports. In a letter to Georgescu-Roegen, Strotz requested his opinions of Ragnar Frisch's suggestions on how to handle the matter. Frisch's letter containing these suggestions is missing from Georgescu-Roegen's files. There is also no record of Strotz's correspondence with Phipps informing him of the ultimate decision. What is known is that Phipps's "Letter to the Editor" was never published in *Econometrica*, nor did its contents ever appear as a note.

From Belief to Knowledge by Proof

At what stage can economists be said to believe that a proof of a proposition in mathematical economics, or economic theory more generally, actually establishes the result that is claimed? When is a proof a proof? The eminent Cambridge mathematician, G. H. Hardy, addressed this question in his 1940 book *A Mathematician's Apology*. Hardy compared a mathematician to an observer gazing at a distant range of mountains. "His object is simply to distinguish clearly and notify to others as many different peaks as he can. . . . When he sees a peak he believes that it is there simply because he sees it. If he wishes someone else to see it, he points to it, either directly or through the chain of summits which led him to recognize it himself. When his pupil also sees it, the research, the argument, the proof is finished" (Hardy 1992 [1940], 17). According to Hardy, a proof is a means of persuasion, it in some part consists of "rhetorical flourishes designed to affect psychology" (ibid.). More generally, of course, the design of proof "to affect psychology" acknowledges the essential social nature of proof, the outward-looking nature of the activity of the proof-maker in attempting to convince another member, or other members, of the disciplinary community, that a particular knowledge claim should be accepted into the community's stock of truths.

In the book that reported the papers given at a conference in West Berlin in 1979, the historian of mathematics Herbert Mehrtens provided an overview of the issues that the historian faces in giving an account of such a process:

> We have to construe mathematics as both a body of knowledge and a field of social practice at the same time. These are not halves of a circular area embedded in the larger area equally divided into science

and society. While the social practice of mathematics is determined by the nature of mathematics as a special type of knowledge, the historical process of extension and change of mathematical knowledge is a social process inseparably embedded in the societal environment. An individual new idea in mathematics is brought forward as a "knowledge claim." This is an act of communications subject to specific social regulations. The evaluation of such a knowledge claim within the community of mathematicians again is a process of social interaction. . . . The inclusion of an interaction into the dogmatized body of taught mathematics, its dissemination into areas of application and other mathematical or scientific sub-disciplines are social processes as well as subject to regulations imposed by norms and institutions. (265)

Mehrtens's point directly touches our own discussion about a major theorem in an applied mathematics discipline, a contribution to the body of mathematical knowledge in economics.

In this examination of the reception of the Arrow-Debreu proof, the community of economics scholars became persuaded of the validity of the proof, thus closing it up in a black box. Did the persuasion occur before Arrow and Debreu submitted the article for publication at *Econometrica*? Both Arrow and Debreu were involved in the internal publication system at Cowles, where their paper quite likely was circulated among the mathematical economists who were members at the time. They presented the paper at the Econometric Society meetings in Chicago with, among others, McKenzie, Koopmans, Beckmann, and Chipman in attendance. It is safe to say that many, if not all, of the most adept mathematical economists were at least somewhat familiar with Arrow and Debreu's proof before it was submitted for publication.

But presentation of the paper to mathematical economists did not necessarily establish the proof's validity beyond all reasonable doubt. To think otherwise is to suggest that the refereeing process in this case or similar cases is merely a confirmation of what everyone (or everyone who matters) already knows. A more forgiving view of the refereeing of the paper is that those involved viewed the process as a means of assessing rigorously the validity of the proof in order to determine whether it would merit wider dissemination among economists. The validity of the proof had to be first accepted by the small community of mathematical economists before gaining acceptance as knowledge among the larger community of economists. Yet the transition from presubmission dissemination of the Arrow-Debreu paper in the small group, to the more open refereeing process, presented a potential problem. The editors at *Econometrica* ostensibly wanted referees who were not biased by previous exposure to the paper, but who were also mathematically adept enough critically to evaluate the paper. These two sets were virtually nonoverlapping. The result was that one referee report emphatically recommended acceptance based on a not very thorough reading of the paper, and another referee report emphatically recommended rejection, reflecting the outdated beliefs of an obscure mathematician. Did, then, this refereeing process achieve its goal of assessing the proof's validity?

The irate response by Phipps to the editor of *Econometrica* raises another question. The ability to persuade someone of the validity of a proof (as suggested by Hardy) rests in part on the mathematical sophistication of the individual being persuaded. In this case we have a paper that is so mathematically complex as to make it difficult for most economists to read it, let alone evaluate it. Yet a mathematician challenged this paper. Are the arbiters to be persuaded by the proof of Arrow and Debreu or by the criticism of the proof by Phipps? And on what grounds are they to be persuaded?

The responses to Phipps's letter point to, among other things, the role of trust in assessing claims to knowledge. The identification of trustworthy agents is essential in assessing and establishing a body of knowledge. As the distinguished sociologist of science Steven Shapin wrote, judging someone's claim to knowledge involves asking, "What are their circumstances and characteristics? What, in general and in this case, do those circumstances and characteristics testify about the likely reliability of what they say?" (Shapin 1994, 38). These, of course, are not the only factors by which individual beliefs become communal knowledge. Yet, in the case of Phipps's challenge to Arrow and Debreu's complex proof, the arbiters frequently contrasted the prestige and mathematical experience of Arrow and Debreu with the lack of prestige and lack of eminence of Phipps. Using the language Shapin (1994), Phipps stood outside the moral economy of truthmakers, and his marginality itself made his view of the proof untrustworthy, and thus finally inconseqential.

Saying this is not to belittle the criticisms of the arbiters who denigrated Phipps's arguments, for they are in good company in their reliance on trust and prestige in assessing a claim to knowledge. Paul Hoffman offered a justification for this reliance on trust in his recounting the life of the mathematician Paul Erdos:

> Today upwards of a quarter million theorems are published a year. . . . But who reads all these theorems? Proof by authority still goes a long way—that you believe a proof because you believe in the person who did the proving or the person who examines the proof. Even Erdos would say "I believe thus-and-such because so-and-so says it's true." Erdos accepted the truth of the four-color map theorem because someone he trusted checked the proof. (Hoffman 1998, 200)

Our look inside the black box of the competitive equilibrium proof has uncovered a very messy process. We have seen that, given the limited number of people qualified to assess the proof, the community of economists was largely persuaded of the proof's correctness by the trustworthiness and distinction of its authors. That the subcommunity was so persuaded was strong enough evidence that the proof was correct that the larger community of economists deemed it to be incontrovertible too. This change in what had been taken to be true knowledge, as knowledge about equation-unknown counting changed to knowledge about fixed-point techniques, took place within a few years of publication of the Arrow-Debreu proof. "Arrow and Debreu have shown that there exists a competitive equilibrium" was black-boxed by the late 1950s. It is then fitting that the summation can be left to one of the protagonists in this story, Kenneth Arrow: "To suggest that the normal processes of scholarship work well on the whole and in the long run is in no way contradictory to the view that the processes of selection and sifting which are essential to the scholarly process are filled with error and sometimes prejudice" (as quoted in Shepherd 1995, vii).

7 Sidney and Hal

I received my Ph.D. in 1955 (note the date) knowing no mathematical economics and recognized that this would soon bar me from reading the current journals. . . . I was on the cusp of the great transformation in modern economics of which I was only very dimly aware.—M. Blaug, "The Formalist Revolution"

I'm enclosing the letters of Hal, dear. I'd like you to hold on to them much as you do mine. I think they are worth preserving. Maybe someday they will give Roy some better insight than school books, histories, written by emotionless unwearied scholars. These breathe life, albeit of a tired weary pen.
—Sidney Weintraub, letter 15 May 1945

Biographical material on Sidney Weintraub is quite typically hagiographic. Following his death in 1983, tributes were written by those with connections to him through the *Journal of Post Keynesian Economics,* which he cofounded with his former student Paul Davidson. More recently three other papers on Weintraub have appeared that have used materials found in the Sidney Weintraub Papers in the Special Collections Library of Duke University. His only published autobiographical piece appeared in the series produced by the Banca Nazionale del Lavoro and was titled "A Jevonian Seditionist: A Mutiny to Enhance the Economic Bounty"? (1983) This was reprinted in Jan Kregel's book *Recollections of Eminent Economists, Volume 1* (1988).[1] The sequence of life events narrated by Weintraub in that paper is reliable, although as is typical, his own interpretations of them are not necessarily what others might provide.

What follows is not the biography of Sidney Weintraub. Neither is it my autobiography as that phrase is usually understood, or his, despite my extended use of Sidney's own written accounts. That is, the present study uses 1) Sidney's previously unknown memorial to his brother Hal;[2] and 2) a set of over a thousand letters that had been in the possession of Sidney's widow Sheila Weintraub, who died on 5 October 1998. Those letters were penned nearly daily during the time Sidney and Sheila were separated from one another first when he was a special student in London before World War II, and then over the years of his military service.[3] During both periods he wrote at least one letter a day to Sheila. Though the letters were primarily private love letters from a student to his fiancée, then a soldier to his wife, they are secondarily a window for scholars looking to reconstruct the training, and interests, of that remarkable generation of economists who took their place as leaders of the American profession in the postwar years.

Sidney Weintraub believed that his 1958 book, *An Approach to the Theory of Income Distribution,* was his most important contribution to economic analysis. That work, developed in articles written over the 1950s, was at that time an unusual attempt to integrate the classical theory of income distribution—the approach of Malthus, Ricardo, Mill, and Marx—with the new Keynesian macroeconomics. That is, it attempted to rehabilitate the classical ideas of distributive shares, or the functional distribution of income, and link them to Keynes's macroeconomic theory of the determination of the aggregate level of income. One needs to remind oneself just how unusual this kind of activity was, at a time when income distribution theory was almost entirely a creature of value theory, a set of discussions of factor pricing and the personal distribution of income. That kind of distribution theory, from Reverend Philip Henry Wicksteed, John Bates Clark, and others, worked in terms of supply and demand curves for particular factor markets. To be sure, some of the British economists following Keynes, like Joan Robinson, Nicholas Kaldor, Piero Sraffa, and others had been raising such issues (and of course Michael Kalecki had made this set of issues fundamental to his analysis) but few others outside the Cambridge group were working in such an area.

Weintraub's book was different. It began with a discussion of Keynesian aggregate supply and aggregate demand determining the level of income that was to be shared across the various groups of wage earners, rentiers, and profit recipients.[4] Weintraub's argument then considered entrepreneurs' expectations of revenues to be received from the employment of workers (and fixed capital) to produce goods to market. Those expectations led to streams of payments to the factors—workers and rentiers—to produce the goods. The factor payments were incomes to the factors, who demanded goods and services, and purchased (or contracted to purchase) newly produced capital goods. Those demands were expressed in the market, producing revenues to the firms. As the revenues exceeded, or fell short of, the revenues expected by the entrepreneurs, the firms shed or hired workers. The Keynesian point of effective demand was that level of employment, and income, at which the expected revenues of the firms were realized by the market outcomes associated with the factor hires instantiated in those expectations.[5]

The book was sufficiently unusual that it could find no publisher.[6] As a result, Weintraub made connection with a small publishing firm in Philadelphia (near the University of Pennsylvania campus), and they agreed to "create" a scholarly book provided that Weintraub took most of the risks and did most of the editorial work. This publishing company, Chilton Company, had been primarily a publisher of auto repair manuals, indeed had nearly cornered the market for such books, but had virtually no experience with scholarly works. Consequently all of the advertising and promotional material was produced by Weintraub himself as he became their entire scholarly marketing operation.

The book carried the dedication "In memory Hal (1923–1954). COE (Bingen), AUS (1942–1946), Ph.D. (Harvard, 1952) Assistant Professor of Mathematics, Tufts College (1953–1954)." And in the last paragraph of the preface, the question of who this person might be is suggested by "I should like to mention the aid given by my late brother on what was for me a difficult hurdle in the chapter on wage theory" (viii).

The following pages will examine in some detail the connection between this well-known economist, and his young mathematician brother. This chapter, as previous chapters, explores the interconnection of the mathematics and economics communities by examining a particular economist and mathematician pair. We will see in this midcentury period, as mathematics and economics were increasingly intertwined, how economists with little mathematical knowledge or training began to understand that intertwining. Part of this story will involve the individuals caught in a time of professional change, but unable to change with the younger professionals. In fiction, this theme was beautifully expressed in McCormmach's characterization of the (circa 1900) German physicist Victor Jakob, loyal to classical mechanics in an emergent age of relativity and quanta (1982). Similarly, our narrative will construct a very unmathematical economist with immense professional ambitions to make a mark in economic theory. But as economic theory was changing rather profoundly at midcentury, such ambitions had to take the form of a yearning for mathematical acceptance. But that acceptance within a community of economic theorists increasingly required a level of mathematical sophistication in economic theory beyond that which would have been considered standard, normal, or appropriate twenty years earlier.

There is a tension between an economics that looks back to history and an economics that looks forward to different degrees of interconnection with mathematics. That this story will be told through the turnings, twistings, and difficulties experienced by a scholar at an Ivy League institution, will lend some credence to our belief, explored over the previous chapters, that telling the story of the development of economics in this century requires attention to the development of economics in its connection with mathematics.

Family History

Sidney Weintraub (no middle name) was born on 28 April 1914 in Brooklyn, New York. His father, Aaron, had immigrated to the United States in 1905. Family stories suggested that he had left the family's home in Upper Silesia in order to escape military service in the czar's army then being mustered to fight the Russo-Japanese War.[7] The family name was not Weintraub, however. Aaron's father's name appears to have been Kummer, and Sidney believed that his paternal grandfather owned a salt mine; Aaron's mother was a Weintraub. After his father's death when Aaron was a young teenager, his mother remarried a man named Bodner with whom Aaron did not get along. Consequently, when Aaron immigrated to the United

States, escaping both czar and stepfather, he appears to have taken advantage of the Ellis Island confusions to enter America as Weintraub. In any event, Aaron's half-siblings in the United States were named Bodner. Sidney's father worked as a stevedore in New York until he had put together enough money to bring over his half-siblings; he then was able to marry and (using his wife's money?) purchase a small grocery store, later a larger family grocery store, in Coney Island. He married Martha Fisch, a second-generation American of German-Jewish descent and they had a first child, William (Bill) in 1912, then Sidney in 1914, then Stelle in 1922, and finally the fourth child, Harold (Hal), in 1923. As often happens in families, over time the children more or less aligned themselves pairwise, with Bill and Stelle becoming close and Sidney and Hal remaining close.

Aaron was an exuberant man, a stocky, burly, and gruff individual who liked to laugh. Once a local rabbi showed up in the store, representing the local Jewish gangsters' protection racket, to suggest that the holdout Aaron pay some money in order that Sidney not be hit by a baseball bat coming home from the baseball field. Family stories then tell of Aaron's coming out from behind the butcher counter with a cleaver and chasing the rabbi down the street. Martha, in contrast, was a dour, sour, and unhappy woman. She believed she had married beneath her, as often was the case for "mixed marriages" between German and Eastern European Jews. Nevertheless, the family prospered and all of the children prospered with them, even through the Great Depression.

We have few records from Sidney's childhood. What accounts he provided to his children were those of a fairly normal city boy, playing baseball in the streets and in the vacant lots, roller skating on the Coney Island boardwalk, and generally participating in the life of the streets of the young immigrant population.

Certainly he was a good student in high school. The evidence for this has to be that he was admitted to New York University from high school at a time when NYU was both a good private university, and not as open to Jewish students as were the city universities. In high school he seemed to concentrate his energies primarily on playing baseball and took a fairly average courseload and course selection, as he was not much interested in becoming anything other than a professional baseball player. His older brother, William (Bill), was earning a law degree, and that kind of professional was well appreciated by the mother at least. Her having married beneath her expectations led her to expect that her children at least would rise above the station to which she had found herself reduced, living above the grocery store on Coney Island, married to a Silesian former stevedore.

In any event, it is clear that Sidney was not a compliant child. His rebelliousness must have been troublesome to the parents who could not find ways to make contact with him in his own interests. If, as all reports confirm, Martha was quite cold and disapproving in general, how could she have found pleasure in a boy who only wanted to play baseball? For Sidney was not a natural academic star. He was not moved along as a precocious academic prodigy, one for whom universities and a scholarly life would appear natural. He played some semi-pro baseball in the summers, and extended his high school career by a year (he called it "redshirting himself") to maintain his eligibility to play baseball. Scrapbooks he kept over all of the years show a good hitter and fielder as a second baseman, and captainship of a very successful James Madison High School team that won the New York City championship in a game played at Ebbets Field.[8]

At NYU he began his studies in the School of Business Administration, enrolling in a Bachelor of Commerce program. This kind of undergraduate business degree is a far cry from present-day business education. It was really a commercial training program, teaching young people the basics of bookkeeping, marketing (once described by Sidney as where to put the lettuce in the supermarket in relation to the bottles of milk), advertising (how to construct an attractive newspaper ad), etc. There was no calculus required in this program, and the only kind of mathematics that Sidney appears to have had was high school geometry combined with some basic algebra. The mathematics course he took in college was what we would now call pre-calculus, involving some advanced algebra and solid geometry.

The stories Sidney told about his college career have a fundamental inconsistency about them. On the one hand, he told about his leaving his dreams of baseball behind because the NYU baseball coach, an Irishman named McCarthy, was an anti-Semite who would simply not play a Jew named Weintraub on a regular basis. On the other hand, he also told a story of his own gradual disillusionment with his athletic skills, based on his very weak throwing arm, which made playing shortstop or third base impossible. And he was not agile enough moving to his right to play second base at this next level beyond high school. Always a confident hitter, his batting was not so dramatically successful that it could overcome the weakness of his throwing arm, and even the minor leagues seemed unattainable. The two stories may in fact converge if the coach made those kinds of judgments, and Sidney internalized them. Nevertheless, by his sophomore year Sidney began to contemplate a life dream ending, and a need to find gainful employment during the Great Depression.

From Second Base to Economics

Sidney Weintraub was not a student who was interested in mathematics, nor did he see mathematics as particularly connected to any of his own concerns. One needs to understand that this is quite different from a number of other Weintraub's contemporaries who became economists. Thinking back on the careers, the educational careers, of Kenneth Arrow, Don Patinkin, and Paul Samuelson for instance, we have a very different set of understandings. For those individuals, mathematics appeared very closely connected to their intellectual interests from an early age. Perhaps none of them was a very competent athlete, or found competition on the playing fields very rewarding. The point is that another field of competition, quite unremarked upon in the history of economics literature, is academic competition for it has a great influence on subsequent scholarly careers.

For many years the community of mathematicians has understood the role of competitive mathematics examinations as connected to some of the magnificent mathematical prodigies—von Neumann, Teller, Szilard, Wigner—who were educated in Hungary in the early decades of the twentieth century. That set of national examinations provided a venue for youthful competitiveness, and it was competitiveness that provided some scope for ambitious and smart high school students to succeed. In chapter 1 we saw Peter Groenewegen, in his biography of Alfred Marshall, make oblique reference to similar matters when he points out that Marshall, after achieving his examination position as Second Wrangler, did not wish to take the prize examination that might have led to a mathematics fellowship at Cambridge, because that next step required some original research. For Marshall, the field of mathematics itself was a competition, a venue for winning prizes. Indeed, mathematics has often been used in that fashion.

Mathematicians and economists are different people. Of course they are communities of people who speak ordinary languages, have social roles, religious and philosophical beliefs, loves, desires, hopes, wishes, beliefs, and fears. We do not deny that they may separately have membership in the community of Roman Catholics, say, or Chicago Cubs fans. Yet to be socialized as an economist is different from being socialized as a mathematician. In the sense that words have specialized meanings, problems have specialized histories, beliefs have specialized justifications, and the networks in which all of these activities take place remain separate and distinct, mathematicians and economists live in different worlds.

In their discussion of the "ideal" mathematician, Philip J. Davis and Reuben Hersch construct a humorous portrait of that person. In a long catalog of characteristics, for instance: "He rests his faith on rigorous proof; he believes that the difference between a correct proof and an incorrect one is an unmistakable and decisive difference. He can think of no condemnation more damning than to say of a student, 'he doesn't even know what a proof is.' Yet he is able to give no coherent explanation of what is meant by rigor, or what is required to make a proof rigorous. In his own work, the line between complete and incomplete proof is always somewhat fuzzy, and often controversial" (Davis and Hersh 1985, 178). Put another way, the utterances of the mathematician are understood quite well by other mathematicians, and hardly understood at all by nonmathematicians. Mathematicians know what they, and their colleagues, are talking about. They know how to be mathematicians. They know what is important.

There really are very few materials on the sociology of mathematics.[9] There were some early attempts to look at the products of mathematics as socially determined, as Marxist and Marxist-influenced sociologists argued that this or that development in mathematics was culturally forced by the particular forms of economic organization (Struik 1942). But leaving such arguments aside, we do not really have a very good understanding of what makes a good mathematician as opposed to what makes a good economist.[10] Who gets trained as a mathematician, and are those people systematically different from the kinds of people who seek training as economists?[11] The folklore of the subject suggests that the mathematician is a misfit, at least in American culture, developing from teenage oddballs who develop a passion for mathematics entirely unreinforced by the larger culture. There is a common belief in departments of mathematics that a mathematician, if not smarter than the everyone else in the university, is at least able with a short period of study to do any other scholarly work that would appear in the university. That belief certainly differentiates the mathematician from the sociologist, who might never suggest, even in the sociology lounge over beer, that the sociologists could do a better job teaching mathematics with a month's preparation than a mathematician could do.

Hubris, overweening pride, characterizes the mathematician's view of his (and until the past decade almost never her) learned profession. An economist can look back to Adam Smith perhaps, and feel a glow and a connection, but the mathematician can claim Euclid, and Archimedes, and Greeks, and Arabs, and Newton, and Galois, and Gauss, and Euler, and Hilbert and von Neumann, and all such geniuses of the past. Compared to these, economists have a past of error and advice to monarchs and claims about a transient social order to their credit—hardly anything, even a pin factory, compares with the prime number theorem. And the Arrow Impossibility Theorem is not the Riemann Hypothesis in depth and complexity. This recognition of genius, and the longing for mathematical immortality connected to having a theorem, or lemma, or inequality named for oneself produces a competitiveness quite beyond what is typically known among academics.[12]

We saw in chapter 1 that the Mathematical Tripos itself developed out of the nature of the Cambridge honors degrees, because mathematics examinations appeared sufficiently objective to allow a ranking of candidates, and thus to provide a formal rank ordering of marks and results in a relatively uncontroversial fashion.

In the United States today, the William Lowell Putnam competition is a testing ground for college student mathematicians, and their mentors. As John Nash's biographer wrote:

> The students that gathered at teatime [in the Princeton mathematics lounge] were as remarkable, in a way, as the faculty. Poor Jews, new emigrants, wealthy foreigners, sons of the working classes, veterans in

their 20s, and teenagers, the students were a diverse as well as a brilliant group. . . . The teas were heaven for the shy, friendless, and socially awkward, a category in which many of these young men belonged. . . . The atmosphere was, however, as competitive as it was friendly. Insults and one upmanship were always major ingredients in teatime banter. The common room was where the young bucks warily sized each other up, bluffed and postured, and locked horns. No culture was more hierarchical than mathematical culture in its precise ranking of individual merit and prestige, yet it was a ranking always in a state of suspense and flux, in which new challenges and scuffles erupted almost daily. There were cliques, mostly based on fields. The clique at the top of the hierarchy was the topology clique . . . then came analysis, . . . then came algebra . . . each clique had its own thoughts about the importance of its subject and its own way of putting the others down. (Nasar 1998, 64–65)

In short, mathematics did permit a number of Jewish students to escape being judged on their performance in either an immigrant vocabulary-laden English language, or with respect to deeper cultural traditions that they did not share. How much more difficult was it to write prize essays in Elizabethan poetry, or reformation history, when one's parents had just arrived, a few years earlier, from Minsk? For mathematicians, mathematics was the same in Poland, Slovenia, or New York. And did not Baruch Spinoza himself construct his ethics in an axiom-theorem-proof fashion?

For Sidney Weintraub, however, the battlefield was not mathematics, but rather second base. And he lost. So from his college junior year, his path was a bit more traditional. He decided that he was going to continue as an economics student, and at least in his last two years began a more intensive pursuit of his coursework. This appears to have led to his coming to the attention of some of the faculty, for whom he volunteered his services as a research assistant. He began working as an assistant to Marcus Nadler, the NYU economics and finance professor, who was writing on international financial markets in the mid-1930s. This research work led to Sidney's lifelong attention to financial markets. For example, this interest led Sidney, in England, to get the occasional check for a financial journalism piece. It also would show up much later, in the 1950s and early 1960s, when he wrote a regular column on the U.S. bond markets, and prognoses for bond market prices, for *Business Scope* magazine, published by Dr. Arnold Soloway of the Harvard Business School.

Sidney received encouragement from other NYU faculty members, and consequently he enrolled in the graduate program in economics at NYU following graduation. It is not clear at what point Sidney decided he wished to become a scholar, as opposed to an economist working in New York at a bank, insurance company, or in the financial press. What he did discuss, in later years, was the fact that he took to the scholarly work, at least the ideas as they were presented in his courses, with real enthusiasm. It was during this time that he did take one undergraduate course in calculus, perhaps feeling a deficiency in his education, although the use of mathematics was not widespread at all in the materials he was reading. He had a long period as a graduate student, working on a part-time basis and supporting himself through the assistance he provided to members of the NYU faculty. And eventually he began a complex doctoral dissertation writing project in economic theory, on the role of monopoly and imperfect competition in a dynamic setting. It was in the process of writing, and attending seminars and workshops, that he made the acquaintance of Tom McManus, then a professor of economics of the College of New Rochelle. The 1938 *Who's Who in the American Economic Association* shows McManus as having received a B.Sc. in 1925 from Northwestern, and an M.A. in 1933 and a Ph.D. in 1934 from the University of Iowa. It identifies his interests as banking and the business cycle, and identifies him as having an interest in economic theory, money and banking, and business cycles. It was McManus's intervention that changed the course of Weintraub's career, for he urged Sidney to seek an academic environment more nourishing than could be found at NYU.

Autodidact at LSE

In his one published autobiographical essay, Weintraub (1989, 40) noted, "Largely under the urgings of McManus, I pooled my meager reserves to attend the London School of Economics, arriving on that famous October 1938 day when Neville Chamberlain alighted from Munich . . . proclaiming peace in our time."[13] Having had a reasonable self-education at NYU, with readings in Marshall, Cassel, and Pareto undertaken under the supervision of Herbert B. Dorau, Weintraub landed in London short on money but with energy, a half-finished doctoral dissertation, and a towering ambition. Writing to his future wife, Sheila Tarlow, he spoke of seeing "Robbins on Friday. He's a tall man, thin hair, very dark combed straight back. . . . He speaks very slowly to make certain of his every word. He outlined a program for me, only advanced work and largely seminars. He said he hoped I'd get tired of the courses after I sat in a few to see how they do things. . . . He also said I'd best work under Kaldor on the thesis. Kaldor is doing most of the theory work at the university. And so it is. As you recall Kaldor is also the fellow who married into the Rothschilds, and to whom I have a letter of introduction. . . . Perhaps I'll have another friend like Tom McManus to talk to" (1 October 1938).

Sidney Weintraub decided to go to LSE in order to complete his education, or at least to rectify the kinds of gaps in his training that he was coming to understand in more and more detail in his frustration while writing his doctoral dissertation with reference to the current literatures. Yet with respect to the hindsight of a future career, LSE in 1938 was probably a poor place to be. The intellectual action, such as it was in the UK, was more centered on Cambridge even though Keynes was restricting his activities following his heart attack. LSE still was in the residual thrall of more traditional classical conservative writers, for whom Hayek and his appointment in the early 1930s had represented a triumph, and a counterweight to Keynes's world at Cambridge even though Kaldor had arrived from Cambridge bringing the message of the new Keynesian revolution to LSE. Weintraub's training then became based on the seminars and workshops in monetary theory and macroeconomics at LSE. This activity represented both the new views coming from Cambridge, and the more traditional microeconomic theory that Weintraub found through Lionel Robbins and Paul Rosenstein-Rodan. But if one sought to maximize one's probability of winning prizes in the future—pace Sidney's ambition, and thus his interest in economics as a "competitive sport"—the optimal move from NYU in 1938 would have been to the University of Chicago. At least there a student or a postdoctoral fellow would have made the acquaintance of all of the members of the Cowles Commission as well members of the Economics

Department of the University of Chicago. A second-best choice might have been Harvard. Harvard was probably out of the question for a special student, for they were not encouraging Jews in the Yard at that time, but Chicago would have been a reasonable choice. In some ways, it is even curious that for one as conservative as Sidney Weintraub was at that time— he did, for example, vote for Alf Landon in 1936—why he did not go to Chicago.[14] It is not that he was unaware of the activity there, for his LSE going-away present, dated 10-20-38, was an inscribed copy of Henry Schultz's *The Theory and Measurement of Demand.*

That there was a sea change in the way economic theory was being presented, and that the connections between the mathematics community and the community of economists was becoming more and more noticeable, cannot be doubted. Some of the difficulties that the Cowles people had with the University of Chicago faculty are a testament to this problem, as was the reception of Paul Samuelson by the Harvard economics faculty chaired by Burbank. However, those issues are all clear in hindsight, for at the time, if Sidney Weintraub is to be representative of what Mark Blaug called the cusp of the revolution, the profession was mostly oblivious to this emerging set of issues. In Great Britain, for example, Keynes referred to *The Review of Economic Studies* as the children's magazine,[15] set up as a counterweight to Keynes's own *Economic Journal,* the establishment voice of the Royal Economic Society. That *The Review of Economic Studies* was not hostile to quantitative and mathematical work, or at least did not share Keynes's ambivalence, is obvious from simple perusal of its contents.

On the subject of mathematics, after arriving at LSE Weintraub wrote to Sheila that "I'm taking a course in advanced calculus which means then I'll have completed all my tool work for economics but German which I will do one of these years. As for the thesis, I've done nothing further so far except I think a bit more of it for discussion with Kaldor then I hope to get busy again and work on it. . . . Robbins suggested I go over to University College to take work under Rosenstein-Rodan, really an advanced thinker and one which I had previously regarded as nearly the most analytical economist anywhere on the basis of his still too few articles available in English. Thus, more theory and I know I'll be satisfied and happy" (9 October 1938).

One of the major issues was that he had gone to LSE while still working on his thesis, and as a result was still formally a graduate student at NYU. The degree would eventually come from NYU though the time at LSE would "count" if he found that he could truly educate himself in economics. Nevertheless, an actual scholarship enabling him to pursue an LSE degree would have been useful. From a letter dated 28 October 1938, he writes, "About the scholarship I doubt that much can be done. I'm not registered for a degree and if I did go after the latter here, it would take a least another year perhaps two more and in the meanwhile I could only expect relatively small grants. . . . Now I think it best I finish at home, NYU, and perhaps a later year, will be back on a post-doctoral scholarship. I think now that I have finally satisfied myself that I am sufficiently 'educated' to want my Ph.D."

Sidney had a lot of difficulties adjusting to life in London. He had very little money, and was not much used to living in such impoverished circumstances. He was trying to write the thesis, chapter by chapter, at the same time he was going to occasional classes and participating in the lively workshops and seminars. For instance, he saved a mimeographed piece dated 27 November 1938, titled "Prospects and Problems of Economic Recovery in the United States," which was presented at the joint Oxford-London-Cambridge Economics Seminar in Cambridge. This discussion looked very concretely at money supplies, and the series of industrial production and prices, gold stocks, etc.

However, it was becoming a problem that as he learned more economics, the kind of economic analysis that he had been doing earlier in New York seemed less and less adequate to handle the analytical problems set forth in the thesis. In a letter of 2 December 1938, he worried that "the only thing I learn here is more and more of my reading inadequacies. I've got to, and will, overcome them. Then perhaps I'll feel lighter and easier . . . it is this which makes my thesis so difficult. I continually hear things from Rosenstein and Kaldor which leave me in a muddle, things that I recognize should come in, but which means that I've got to change my own attack somewhat. But do remind me that a doctorate hinges on finishing it. Refinements can come later. Yet I don't relish working when I'm not certain in my own mind."

Those kinds of uncertainties were to surface all during that year in London, though they ebbed in their intensity: after a few months he could say

"I feel so much better. Chapter two will be ready for delivery tomorrow or Friday, chapter three the following week or ten days. Also I've learned a great deal since a little while back. Probably the things I've learned are old, but yet I'm learning them so I don't mind. The thesis I think will go better in the future. I still need some urging. . . . I must finish it . . . so that I can get that damn fraud—a degree" (21 December 1938).

One of the features of studying in London, and seeing the English professional landscape in economics, involved comparing it to his own experience, and employment limitations, in the United States. He was especially struck by the large number of Jewish faculty members he met. In a letter dated 21 December 1938, he comments on the "craziness" of Mussolini throwing the Jews out of all university posts in Italy. He wrote that "Piero Sraffa, Professor of Economics at Cambridge with whom I had lunch . . . is Jewish I have since learned.[16] He is an outstanding figure, one of the very best. He had to leave in 1927 because of [his] anti-fascist activity. Keynes brought him to Cambridge where he has remained since. . . . Here at London, without thinking, look at the professors: Ginsberg, Mannheim, Laski, Rosenstein, Kaldor, Lerner, and I'm sure that does not exhaust the list."

Sidney was to leave LSE with the thesis essentially finished, and with an exceptional command of the past literature in economics. But except for the single mention of the course in Advanced Calculus, which he appears never to have attended, his reports on his studies only refer to repairing gaps in his education with respect to the kind of work being done in economic theory in England at that time, and such training did not include mathematics. His teachers like Kaldor, Robbins, and Rosenstein-Rodan were not part of the mathematical wave about to break over the profession. Of J. R. Hicks, of R. D. G. Allen, there is no mention. Sidney Weintraub had invested heavily, at his own expense, in finishing his graduate training in economics at LSE. This experience was to be the touchstone for his counsel to his future students, namely to find a great library and read oneself into the profession. But since he was not one who easily could tolerate instruction by others, nor could he easily tolerate competing on a playing field like mathematics in which he had no advantage, his studies included no mathematics. In retrospect, he had missed his chance to become the kind of important economic theorist he so wished to become.

Returning from the London School of Economics in the summer of 1939, he needed to find a job. The Depression, though it was winding down, still had little room for academic economists, or private sector economists. Washington, of course, was always an option, one that he would eventually take up. But the issue of finding an academic position, one that would allow him to write, to make an impact as a scholar, was the real challenge. After all, that is why he took the financial risk to go to LSE. But starting a professional career, following the year at LSE, was not to be easy. He wrote to Sheila that "[Tom McManus] had seen [Dean of the Business School at St. Johns College] Weary and the latter would like to hire me. However to do so he would have to explain my religion away to the President's satisfaction. Thus rather than go to this trouble he was first going to interview three Christian boys. He would however soon let me know. Then he saw Steiner. Steiner said he thought perhaps I was the best man but I wasn't too responsive to the offer. On the other hand the others interviewed were eager to accept it. Tom pointed out a little difference: that I'm five years out of school and they either a year or two" (1 August 1939).

Sidney was on an outsider's path. It was not the case that he had gone from a distinguished undergraduate career to the kind of exceptional graduate program at Chicago, or Harvard, that was to define the future intellectual course of many of his contemporaries whom we now have honored with Nobel prizes. NYU was nowhere. LSE was, however, somewhere. How then was he to find a permanent job? All of these concerns, and anti-Semitism too.

The actual solution Weintraub found was driven by external events. After a brief spell in New York, teaching on a temporary appointment at St. John's University, immediately after Pearl Harbor at Walter Salant's invitation he went to Washington, D.C., to work at the U.S. Treasury. He quickly moved to the Office of Price Administration, and disliked that work too. He then moved back to New York in 1942 and took a position at the New York Federal Reserve Bank editing its *Monthly Review,* and tried to get a commission in the military. His two brothers were in the service, and although he was at that time draft-exempt as a twenty-nine-year-old, his competitive nature would not permit his remaining a civilian. Failing to get an officer's commission, Sidney enlisted in the army, and he reported for duty in March, on the day after I, his first child, was born.

An Economist at War with the Army

There are numerous autobiographical accounts of economists who came of age professionally in the immediate postwar period, and who had had a "good war." That is, the accounts often refer to the heady environment of the Statistical Research Group (Friedman, Stigler, Wald, Wallis), the Office of Strategic Services (Roosa, Hoover, Galbraith, Rostow, Salant, Barnett, Kindleberger), the Combined Shipping Board (Koopmans), or other work like meteorology (Arrow), Ordinance Laboratories (Samuelson), service as naval officers like Bronfenbrenner, Buchanan, and Tobin. In this language, a good war involved a real contribution to the war effort, a sense of participation with like-minded others in interesting or important work, and a belief that the personal contacts made were useful for their future intellectual or professional growth and development. Sidney had, in this language, a very bad war.

In retrospect, the rebelliousness that was to lead him to call his autobiographical self a "Jevonian Seditionist" had been present in his short time at the Treasury, O.P.A., and the New York Federal Reserve Bank. It certainly was manifest in his refusal to act in subordinate roles and to take direction from others whom he believed had less ability than he. But behaving as a "Seditionist" was not exactly conducive to a successful army career. Sent to the Army Quartermaster Corps, Sidney twice went through basic training, and suffered basic training for his clerk duties three times. He was turned down for Officer Candidate School several times by his immediate superior officers with whom he conducted his own private war, spending his time reading, writing letters, and disappearing for many hours at a time on ten-minute assignments. His superiors retaliated in usual course by keeping him on K.P., latrine duty, and in the casual labor pool.

Consequently, he had time in London, as the Allies made increasing progress in the war against Germany, to begin to reconstruct a possible postwar future for himself as an economist. While on "detached service" with an intelligence unit in London (he was a file clerk, and spent his time playing softball in Hyde Park), he wrote to Sheila:

> I saw Tibor Scitovsky today, and in a role in which he looked so uncomfortably out of place: he was on KP . . . I just had snatches of

conversation with him. I still did not, by any means, regret my lassitude and failure to make more strenuous efforts to join his unit. I do not relish the thought of full days in the library, on subjects detestable to me, learning of enemy industries, jurisprudence, geography, etc. all for the compensation of EM [enlisted man] promotion. I just am not interested in writing for army incompetents or showing them anew that I am a professional economist. They can go to hell. If they wish to use me as such I want commensurable [officer] status immediately rather than promises which like as not would turn out to be in vain. Moreover I would not want to work, I think, out of personal loyalties solely to the friend of mine who happened to be in charge. The organization does not hold that much intangible attraction to me. (15 February 1945)

Sidney did have an interesting solution to the boredom of army life, and to his belief that he was simply wasting his time in an army that would not utilize his talents in productive ways. He decided not only to read all the kinds of books he never had time for as a student, but also to study mathematics. The book he carried around from assignment to assignment was *Advanced Calculus* by William F. Osgood (1943): "This afternoon I had every intention of studying some math. But these last two nights have been so choked with work that I have been too exhausted to do so; I had actually to nap. I'll try again tomorrow. It's almost utterly dark here now" (1 November 1944). "Actually dear, last night I worked in that advanced calculus book for the first time in about two months. I believe now I may be able to keep at it regularly. I will at least try to do so" (22 November 1944). "Now for this evening. I'm studying math again and learning French from a new larger grammar. I'm really progressing fine" (8 December 1944). "Last night I actually returned to the advanced calculus volume and made some progress, starting at the beginning and working problems. I intend to proceed as religiously as possible in the circumstances hereafter. I also read more than half way through Lytton Strachey's *Eminent Victorians*" (18 December 1944). "Well today just a routine day but I did manage to start my math tonight. Yes, honest to goodness dearest. Henceforth that will be a nocturnal ritual 'til they ship my carcass home, when I can once more resume my studies disturbed only by you and Roy" (26 February 1945). "It's

Figure 1   Grandfather Aaron (far right) with wife Martha (others unidentified) in front of their Coney Island grocery store. Note roller coaster reflected in store windows.

Figure 3   Sidney in Hyde Park, while at LSE, 1938.

Figure 2   Sidney, the baseball player, 1931.

Figure 4 Sidney at war with the U.S. Army, High Wycombe, England, 1944.

Figure 5 Brothers Hal (left) and Bill (right), on furlough in London, 1944.

Figure 6 The author with economist father (left) and mathematician uncle (right), 1953.

11 P.M. now [and] from 8 to 10:45 I have been doing math, in Osgood. I've made a vow not to skip the deductive problems nor to go on when I fail, through close reading, to understand. I mean to mark the book or at least recognize those portions which I deem unworthy of concentrated study. I also intend to buy in town, shortly, another math volume which is both simpler and more appealing to the reader. Also I want to pick up a cheap edition of *The Education of Henry Adams,* a work I've meant to read for some years now" (5 March 1945).

The issue was reasonably clear. Sidney intended to return to the academic world, and his own writing was going to be in economic theory. He had been planning to turn his unwieldy dissertation into a book, and had completed some of that work prior to his induction. He had given the manuscript to Fritz Machlup at Princeton for comments, but Machlup was unresponsive.[17] The only course appeared to use the war years as a time to learn more, and more appropriately, for the economist's life in the future. For Sidney Weintraub, that meant the study of mathematics, for he felt his inadequacies in this area acutely, and seemed to appreciate that mathematical skills were to be necessary for a serious career in academic economics.

His reach, though, appeared to exceed his grasp. Osgood was a mountain to climb, but it did not yield to the kinds of approaches to learning that Sidney found cognitively congenial. All through the fall and winter of 1944 in France, and thereafter through 1945 in London,[18] in letters to his wife Sidney recounted the struggle to understand calculus: "Also, on lunch hour I went into a bookshop once more, thumbed through some familiar relics to attempt to get copies of a math book I want, Hardy's *Pure Math[ematics].* . . . About math, I'm becoming obsessed. I do intend to master Osgood's *Advanced Calculus* or I ain't acoming home to you" (6 March 1945). "It's 11 P.M. now and I'm actually getting tired. Since 7:00 I've been sitting here, 'til this moment doing some math. Believe me? Yep, I'm intent this time to do more than turn pages. During the day I try to recall sections and topics discussed by the author, to fix at least the method in mind. As success crowns my endeavors, I'll finish this in about two months and then I've been thorough Osgood and advanced calculus, a milestone in my vainglorious, obscure achievements while someday, somewhere, other than the personal edification, it may stand me in good stead" (7 March

1945). "Just to show how completely this strain of extravagance overtook me, I went to town to purchase a volume of Hardy's *Pure Math[ematics]*. It may be a month, or even a little longer, before I can get to it. But as a gesture, I wanted to make a start, to secure possession of it. So tonight I'm sole owner. As a matter of fact I've been studying, or at least glancing and trying to fathom, some of Osgood for some 4 hours now, a really good evening and turn at it. Progress is pretty good. Maybe I will become a math student after all. But I do have a long row to hoe" (9 March 1945).

One needs to recall that Sidney would not have been permitted to take a course in Advanced Calculus were he to have been a regular college student. The usual process was that two years, or four semesters, of first elementary then intermediate calculus led up to a course in "Advanced Calculus." Sidney had had but one semester of calculus at NYU, and a visit to an advanced calculus class at LSE. He was not well-prepared, and no amount of reading would repair those deficits. Sidney was the kind of math student who drives calculus teachers to despair. He looked over the problems in each chapter, and did not do those that he found too easy, and skipped over those he found too difficult. Calculus teachers insist that such students ask questions, and work with others to solve the problems, for no amount of self-study of a basic text will allow a student with only average training to master the material by solving the textbook problems.

Thus Sidney's recounting of his own educational adventure reads like a kind of war story, a forced march to a distant objective through an enemy army called "Osgood": "Today wasn't without progress. I learned all about Lagrange multipliers, something which had stymied me and recurred time and again in mathematics-economic analysis. I've made it part of me now" (16 March 1945). "I've an opportunity to master a new tool and mark a new achievement in math while others are too busy to learn. Rightly or wrongly, I want to do it. Then, if I'm still cut out for it, back to social questions. Likewise I want to read to recompense for the years of denial" (26 March 1945). "By the way, in my going through my math, you know I've been doing it at practically a gallop, covering in the environs of a hundred pages a week. By the end of next week, I would have completed the book, at least this reading. Thereafter, I'll go through once more, at a more pedestrian pace of 25 pages a week, or about 5 an evening. So long as I now have some of the general notions I can afford to master all of them. As

I have the time I can do the less ambitious project and task even if it implies so much longer before I obtain a sufficient degree of mastery of the subject. That, you know, is the trouble with me and my stage, the impatience to undergo the pages, the slow accumulative process of learning, anxious to obtain the same level of proficiency as anyone's main sphere. But considering that three to four months is not too much of a price to pay for an accomplishment which to adhere for a lifetime, I'm determined to do it. Simultaneously with my new reading of Osgood, I'll proceed to go through Hardy at a gallop. Before I do all that however I freely intend to intersperse the time with lighter reading, completing most of those Modern Library editions which I've collected in almost enormous quantities" (2 April 1945). "You then compliment me on my self-imposed discipline, my pursuit of math. Sometimes I too am amazed at my own persistence. Fundamentally it probably suggests, as I told you yesterday, that I can do no other, that it was inclinations run that way whatever the time or place. Though I should have more than my feel of Osgood by this time, I continue to adhere to it. I shall master it. You know, apart from more in the way of a literary background and its parlance for writing it was the only tool, the only bit of mental equipment I lacked. So I propose to remedy that under the circumstances" (13 April 1945).

Sidney had enough demobilization points to return to his family on Christmas Eve, 1945. As an economist, he had had a bad war. His talents unutilized, his energy spent avoiding work assignments like collecting garbage and sweeping officers' billets and cleaning their toilets, he had at least managed to survive physically despite his attempts, in 1944, to get assigned to a line rifle company. The army felt his services as a clerk were invaluable, and so he remained a clerk for the duration with the occasional brief stint as a mailman, chauffeur, POW translator, or mail censor. He had tried on his own to learn mathematics, believing it necessary for his future work, but the task was really too hard. Nevertheless, for him the real answer to the value of his wartime service was that "it hasn't all been a waste of time; it helped me somewhat to learn, to see what makes people tick, to count the all too few honorable specimens. I've realized much of this before but seldom had I imagined so many were stooped, stupid and mired. Maybe too it's been a refreshing sight in a way, a rest from economic to individual analysis" (26 November 1944).

Introducing Hal

> Stelle was just one year old
> Hal was born. It was 1923.
> He was a scrawny babe
> Sickly, big-limbed, almost deformed
> One aunt advised against keeping him
> They all gathered on a Sunday,
> Shortly after his birth.

Compared to Sidney, we have much less information on Harold Wein-traub. Some of this results from his extremely premature death. He was nearly nine years younger than Sidney, though their closeness seems to have developed early, as the older boy was the one to socialize the younger into the arcanae of sports and street life.

> Hal had to learn to take teasing.
> He got it from Bill and myself.
> We called him "Stosh" as we
> jostled him about.
> Most vividly I recall, about 1929,
> Hal was six when I put a small
> baseball bat in his hands.
> I made him hold it left-handed.
> Compelled him to swing evenly.
> Reprimanded him for lurching.
> We played in the alley. He hit
> several balls good. Way out into
> the street. Suddenly he cried.
> I insisted on perfection.
> I forgot he was a small boy. He resisted.
> I wheedled him into returning. He continued
> batting. To the end
> he could only bat a ball left-handed.

Though immensely successful in school, Hal was young and immature for his class cohort, over protected apparently both by his mother and by

Sidney, his cherished older brother. The age difference is important. When Sidney was in college, Hal was still a child.

> In 1931 I played high school baseball. Hal,
> in school now, accompanied me when he could.
> In 1932 I played with the college freshman team.
> Hal came along on the bus. The trip was against
> Concordia Prep. in Bronxville, I think.
> Hal saw an animal. Excitedly, "What's that?"
> Alas, city dweller, it was an unsaddled horse.
> He had been skipped several times.
> Came 1935 he
> graduated grade school, I college. He won
> practically all the prizes for merit and
> proficiency. Mom and Pop welled.
> This boy made them proudest of all.

Hal was a young high school student, age fifteen, when Sidney went off to LSE in 1938. Hal thus had, in contrast to brother Bill, a lawyer at that point, one member of the family who had already made some decisions about a scholar's career.

> I was now an ardent student. Hal and Stelle
> became bookworms. Hal was the more successful.
> Stelle, poor girl, could not keep the pace.
> I thought he showed amazing promise for
> development. Apparently he was too young
> for his school classes.
> By 1939 he was ready to graduate.
> Again, all the honors heaped upon him at
> Abraham Lincoln High School.
> This boy, destined to be a mental genius.
> All of us proud, reassured, confirmed in
> our beliefs.

He entered Brooklyn College, but of course the war intervened. There is some evidence that he was studying mathematics at Brooklyn College before he was drafted, and that he had combined that study of mathematics with some study of physics, but of course in college one is not a specialist, nor can one be said to have exceptionally well-developed interests. In the letters that Sidney and Sheila exchanged, there is some discussion about Hal's considering engineering as a possible future. Perhaps Hal was influenced by Stelle's engagement to an engineer. In any event, Hal was drafted at age twenty and his college career at Brooklyn came to an end.

> In February [1943] Hal was drafted, tho he was a
> senior, ready to graduate. Under my prodding
> he had volunteered for the Air Corps, as a
> Flying Cadet. I thought it would be a
> cleaner life as a commissioned officer.
> I recall his failing the Cadet program, on
> grounds of eyesight and, more so, lack of
> enthusiasm for it.

We can track some of Hal's educational changes over the next few years as he was sent by the Army Corps of Engineers first to Lehigh University, then to Grinnell College in Iowa (where he got the mumps, and had to return to do basic training a second time), and then to the University of New Hampshire where he to received his B.S. degree in an accelerated wartime program. "How do you like Hal's comments on New Hampshire? He likes the good things of life doesn't he? I think he's safe for the future, not a lost mind who will wonder what to do" (22 October 1944).

> He wound up at Jefferson
> Barracks, outside of St. Louis, doing permanent
> KP. Here he threatened dire things, that this
> was inhumane, that a daily sixteen hour kitchen
> stint was unendurable. Patience, patience,
> was all I could counsel.
> Then, an opportunity. Under the ASTP program
> the bright youths could get advanced college
> training in engineering. Hal was accepted.

Nevertheless his war experience reflected some of Sidney's own. His letters from the army to Sidney, and Sidney's recounting of them to wife Sheila, speak of a near desperate disenchantment with military life. This was combined with Sidney's own fear that his extremely talented and favorite brother was going to end up sweeping mines on contested roads in France and Germany.

"So Hal has shipped over too. That's three of us, 100 percent, except for Pop. Well, that's about all that can be said, I'm sure. Given at all a chance he'll succeed but I'm so dubious about the army and their desires to ferret out ability. Aggressiveness not acuteness is what manages desire, among all ranks. And it never was an ameliorative human quality" (11 November 1944). "Hal wrote. He has been reading Rolland's *Jean Cristophe* and was struck by the human parallels, a projection of even his own experiences. He enjoyed it. I must read it as soon as possible" (13 December 1944). Hal wrote of several of his ASTP friends already being used to dig mines ahead of the 7th Army in Alsace. "Hal's own training program nowadays is rigorous and he is much too exhausted to write" (2 February 1945).

Though Hal thought he was going to be trained for a technical-scientific position with the engineers, it became clear very quickly that their group was destined for the lines, and that carrying a rifle and using it was going to be part of his military career. The tension in the letters Sidney sent to Sheila, and which crossed with Hal's own letters, is palpable until it began to emerge that Hal's unit was operating just behind the front line, primarily engaged in actual construction of bridges across rivers in France and Germany. "So the youngster's [Hal] a dough[boy] with the Third Army, having helped bridge the Rhine near Cologne. You probably have later news; mine was dated the 9th of April" (29 April 1945).

> Home for me, December 1945. Hal still overseas.
> Attending school at Nancy (?), France, delighting
> his instructor with his math acumen.
> Rejecting an offer to stay on a fellowship.
> About March or April 1946. Home again for Hal.
> Happy days. We saw him frequently at Greenwich
> Village. He loved our apartment and our company.
> I never could get enough of him.
> I think he attended, and taught, as a tutor at
> Brooklyn College that summer. His confidence
> returned at this opportunity of earning a living.

How abysmal his army days had been. Nary a
promotion, never a recognition of talent, not an
opportunity to issue an order, however trivial.
Now he was picking up the mathematical threads.
He had been the star math major prior to his
induction. He had been President of the Brooklyn
College Math Club. I wanted a math career for
him, at Harvard. Sister Stelle, and her husband,
half-persuaded him to the virtues and earnings,
in physics and engineering. He was tempted to follow.
My views triumphed. He did attend Harvard tho
accepted at MIT. He received an MA quickly, then
invited to be a teaching fellow. He accepted,
flattered. He loved his days at Harvard, the contacts.

There is a gap in the written record following the war, but it appears that Hal went to Harvard on the GI Bill, as a Ph.D. student in mathematics. Sidney played a major role in counseling this course, as Sidney by that time had seen the differential reward associated with pursuit of mathematics compared to economics, and it was, of course, the post-Manhattan project period in which scientists were pushing out what Vanniver Bush called "the endless frontier." Sidney was not disinterested in Hal's own academic success as it might reinforce and validate the choices he himself had made. The Weintraub family conflict over Hal's future was one of the defining issues in what ultimately led to the breakdown in the relationship between Sidney and Stelle. It was related it seems to Sidney's belief that Stelle was overly concerned with material things, and insufficiently accepting of those who wished to follow a different, more cerebral, path. Stelle, her mother's daughter, was thought by Sidney to judge human worth by the quantity of consumer goods they commanded. From their mother's perspective though Stelle, unlike Sidney, had married well, and Bill was at least a lawyer. For their mother Sidney had married East European trash of no social significance, and Hal was in danger of following in Sidney's footsteps. Thus the issue of Hal's becoming a mathematician was hedged all around by the desires, beliefs, and cultural understandings of the difference between a scholar's life, and a commercially successful one. And thus for Sidney, Hal's career projected his own wishes for himself onto his beloved younger brother.

So Hal went off to Harvard to do his Ph.D. I have no records of his graduate work, in the sense of his course work or his performance as a graduate student prior to his dissertation. One difficulty is that there is a divergence between family stories of Hal, the exceptional genius, brilliant mathematician, whose life was cut tragically short before he could do the major work in mathematics, and the reality of his academic success, which was not great.

> In his third year at Harvard, striving for his Doctorate,
> Hal failed to achieve it. He was unsure of his ability,
> or his topic. I prodded him compelled him to go on.
> He did.

Aaron Weintraub died in December of 1949, and the following May a very sick Hal was diagnosed with Hodgkin's Disease.

> When I
> visited Hal in January 1951 I also saw his dissertation
> supervisor Professor Ahlfors. I told him of Hal's
> physical condition, inquired whether I should encourage
> him to seek the degree. The reply was in the affirmative,
> explaining that Hal had some ill-luck in the
> choice of a topic and that he (Ahlfors) had not
> been as free with his time as he might be. He
> promised to rectify it. Hal did get his degree,
> strictly on merit I am sure, that summer. Mom
> and Bill went up for the occasion, and rushed
> right back, with Bill insisting on driving Hal's car!

We have various letters that have suggested that Hal himself was not a particular favorite of his thesis advisor, the eminent Lars V. Ahlfors. Ahlfors did not really seem to have so much enthusiasm or confidence in Hal that he was willing to put him on the job market at the dissertation stage. Hal's letters describe the process of being frustrated, of not making much progress, and of not having very much contact or encouragement. Of course some of this is the usual plight of a graduate student, but there does in fact seem to be somewhat more involved here. His doctoral dissertation "Borel Monogenic Functions" was signed by Ahlfors and David V. Widder who probably formed his committee. It is forty-six pages long, and is dated 1951. The Harvard archives suggest that it was checked out, as a thesis volume, three times, all between July 1951 and December 1952. The dissertation is in the field of complex variable theory, complex analysis, and its structure is associated with defining a class of functions, Borel monogenic functions (complex valued functions of a certain type) and proving several theorems similar to theorems already in existence for more tightly defined functions. That is, Borel monogenic functions are one sort of generalization of analytic functions (functions expressible by power series expansions). Some classic theorems concerning analytic functions have analogs for Borel monogenic functions, and Hal's thesis proved several theorems of this sort.

On Ahlfors's recommendation, Hal was able to get a teaching position at Tufts College, and was hired on the regular faculty there as an assistant professor when he received his Ph.D. It was there that Sidney encouraged Hal's doing work in economics, and sought his help with mathematical formulations of problems that Sidney had found in his own work in economic theory.

Collaboration with a Brother

"Returning home, bereft of a job, having a few contacts, it was McManus who performed the pivotal rescue with a teaching post at St. John's University where I worked for the most decent supervisory Dean whom I was ever to encounter" (Weintraub 1988, 45). St. John's was then located in Brooklyn, not too very far from where Sheila had been living during the war. The kind of work he did there during that G.I. Bill period of high college enrollments, and few teachers, might make the current professorate uncomfortable:

"I believe I have my tentative program for next time. Monday, Wednesday, 9:00 to 12:00 with two courses in Economic Theory. Tuesday, Thursday, 12:00 to 1:30 in Economic Theory; my guess is, however, for the dean said as much, I'll have classes Tuesday, Thursday from 9:00 to 12:00 in Econ

I. I won't mind so long as I have to come in for the noon class in Theory. Then Tuesday, Thursday evenings, 6:30 to 9:30, one course in Theory, the other is a new one, in Advanced Money and Banking. In all, I think, four courses in Theory, two in Econ I and one in Banking. This will make 21 [credits] and should go well over $500 per month or, including summer school over $6,000 per annum. It could be worse though it is a heavy load . . . tomorrow chapter sixteen should be a memory. Maybe the next few will be easier" (25 July 1947).

It was a difficult period, although compared to the army service it was a heaven of autonomy. He returned to his revised thesis manuscript, and worked at it continuously.

After *Price Theory* was finally published in 1949, and with *Income and Employment Analysis* written and ready to appear in 1951, he sought another position, and turned down the offer of a job at Indiana University to take an appointment at the University of Pennsylvania.[19] What is surprising, though, is that he was recruited to teach not only economic theory but also mathematical economics at this Ivy League institution.

"Dear Professor Bowman . . . I would have no objection to direct inquiry by you of my immediate supervisor at St. Johns, Dean William Weary, if you are prepared to make a firm commitment. Recalling our conversation, the idea of doing a course in Mathematical Economics appeals to me. I would be ready for it by the fall and would also be willing to specialize and branch out in that direction" (3 January 1950). "Dear Professor Weintraub: I am extremely happy to offer you a visiting professorship in Economics for the academic year 1950–51, at a salary of $6,500 for the ten months. As I explained in previous conversations, we have planned three courses for you to teach; namely Economics 611, Introduction to Mathematical Economics, Economics 604, Recent Developments in Economic Theory, and Economics 605, Seminar and Selected Problems of Economic Theory. . . . Sincerely R. T. Bowman, Chairman, Department of Economics, University of Pennsylvania" (18April 1950). "Dear Professor Bowman: Thanks very much for your letter of April 18th. You may construe this as my acceptance of the offer outlined in the letter" (20 April 1950). "Dear Professor Bowman: Enclosed are some indications of what I have in mind for the three courses . . . all of the course plans are fluid enough I think to permit some variation in the light of developments in the literature of classroom needs.

For example, it is my impression that Tintner will have a book on Mathematical Economics available next year. If so, I may use it toward the close of the course" (13 June 1950). "Dear Professor Weintraub: . . . the course materials in your outline seem very excellent to me. The material in Mathematical Economics is exactly what I had in mind"(from Raymond T. Bowman, 26 June 1950). "Dear Professor Weintraub: I am extremely happy to be able to offer you a full professorship at the University of Pennsylvania, beginning September 15, 1951. The compensation approved by the administration is $6,500 per year. . . . Sincerely, R. T. Bowman, Chairman" (10 April 1951).

Sidney Weintraub spent his first year at Penn, 1950–51, as a visiting professor, with a verbal assurance that it would be converted to a regular position in the following year. This offer finally materialized in April, as the tenure decision was made by that time. It was fortunate that that happened, as Sidney had already moved his family from Brooklyn by then, and had taken up residence in the suburbs of Philadelphia.

> In August [1951] I visited.
> Hal and I went to Cape Cod. Now he loved to drive,
> a little recklessly to show me he could handle the
> wheel. We spent 3 friendly, intimate days together,
> full of companionship, compassion, in spiritual and
> filial warmth. These were to be among the happiest
> days of my life. I left Hal exhilarated by the
> contact, downcast at the symptoms of chronic ill-
> health tho this was a period of reasonable strength.

It was in this period, from the summer of 1951 until Hal's death in November 1954, which saw Sidney and Hal begin to form an adult relationship that involved the multiple elements of adult brothers, scholars at different universities, economist and mathematician, and mentor with pupil. It was also made more intense by their mutual understanding of Hal's inevitably fatal cancer.

Not very many of Hal's letters have survived; there are few letters except for what his widow might possess or have kept. The first letter to Sidney is dated 14 February 1952, and begins by thanking Sidney for his help at Hal's wedding.[20] "Mentally I was terribly depressed and you were largely responsible for pulling me through. You always seem to be near by, and helpful in the large moments in my life." Then going on to talk about how he and Natalie were establishing a life as a married couple, he wrote, "This term Clarkson (the Math Department's Chair at Tufts) has given me an especially light schedule. I have three courses which take ten hours; I have the course in complex variables and two elementary courses, the advanced course has but six students and so far has been going well. When I looked surprised at the abbreviated schedule, Clarkson remarked, 'Well Hal you are becoming a big wheel.'"

Then, in what is quite a suggestive remark about the kind of research mathematician Hal was, or wasn't, we find: "Sid, I do have free time, and would be very interested if you had any suggestions for an article. I know that you spoke of possibilities before, but I would mainly like to know the sort of reading I would have to do, and what questions would be posed and answered. I would like to finally create, but on my own in mathematics I don't seem to be overly imaginative. I've been looking at journal articles and none of them interest me enough to try to add to them. Of course I've just started. But I would appreciate suggestions."

It is instructive to see this questioning of Sidney by Hal concerning what kinds of work in economics a mathematician might direct energies. Compare this to the George Miller mathematics dissertation (chapters 5 and 6) turned in at the University of Florida in 1951, at nearly the exact same time that Hal turned in his own doctoral dissertation in mathematics. Compare Miller's acceptance of Phipps's understanding of economics with Hal's view of Sidney. The contrast between these two pairs of mathematicians and economists is interesting for while Miller's mentor could at least be said to have known some mathematics, while maintaining a pristine ignorance of economics, Hal's brother was exceptionally well-read in economics, and quite oblivious of the kinds of mathematical work just being constructed in the discipline. This gives a fair reading of the state of what was then called the community of mathematical economists in the 1940s. They were like feminist theorists in the 1970s: "They were eccentric because they had failed to fit into roles that men had contrived for them to fill and because there were as yet no other roles. For roles require a community—a web of social expectations and habits that defined the role in question. The community may be small, but like a club as opposed to a convocation, or a new species as opposed to a few atypical new members of an old species, it exists only insofar as it is self-sustaining and self-reproducing" (Rorty 1998, 226).

In any event Hal's illness prevented such future work. What we do have, however, are a few letters written a bit later that connect to material that surfaced in Sidney's 1958 book, or at least the articles that lead up to that book. On 9 October 1953, Hal wrote:

> Here is an attempt at an answer to your queries. (1) In reference to question one. If $N=N(Z)$, $Z=Z(N)$ is called the inverse function on $N=N(Z)$. [N. B. by definition a *function* $N=N(Z)$ is single valued, that is for each Z there is only one N; thus the inverse function may not exist; that is for some $N=N_1$, there may be more than one Z (say Z_1, Z_2, Z_3); see figure. The inverse function will exist if N increases with increasing Z or N decreases with increases Z. (2) [In reference to question two.] Parametric equation idea. A curve is represented parametrically by two equations $x=x(t)$, $y=y(t)$ where the parameter t is given real values, x and y are thereby computed and plotted in x, y plane to give curve in xy plane. (3) Given two curves represented parametrically. . . . etc. (4) Now to your problem. $D=D(N,w)$, $Z=Z(N,w)$. For each w we get two curves, then varying w gives two families of curves; fixing w [enables us] to find point of intersection of D,Z curves. Put $D(n,w)=Z(N,w)$. This permits us to solve for coordinate N of point of intersection in terms of W, say $N=\Phi(w)$.

This discussion then goes on to derive what effectively are supply curves for labor. The letter ends, "I can't give you a specific reference. Probably Courant has a good section on parametric representation of curves."

We have the letter Sidney sent to Hal with the questions, and that was Hal's reply. There are two features of this exchange that are worth noting. First, Hal's "solution" to Sidney's substantive question appears nearly as is in the "Appendix" to chapter 6 of Weintraub's 1958 book, and underlies the article he published in the *American Economic Review* of December 1956 on which that chapter was based. Nevertheless, comparing Weintraub's treatment of that chapter's topic, labor demand and supply functions, with other work done at roughly the same time by others (Arrow and Debreu, for instance), it is painful to read twenty-three pages of analysis of mathematical functions with so little mathematics actually employed. But this leads to the second curious feature of the exchange, namely how elementary the issue is, really. This kind of discussion of parametric equations is part of an intermediate calculus course, and is well below the level in which even advanced calculus material would be presented to what are now college sophomore or juniors. The reference to Courant's book is to *Differential and Integral Calculus* (1937 [1934]), in two volumes, which was the canonical reference work for the calculus in the United States in the earlier period of the middle 1930s through the 1940s. Indeed, the first of the two-volume set in my possession is inscribed "To Hal, and don't speak mathematics to me until you're through this. Sid. 11/27/40." The second volume is inscribed "To Hal, a fair exchange—with much confidence. Sid. 11/27/40." Thus Hal's referring Sidney to Courant to look up the material on parametric equations is a small irony since these volumes were those that Sidney had in fact given Hal fourteen years earlier.

Another letter, 10 April 1954, appears as well to have been constructed at least in part as a reply to a set of questions that Sidney had raised about the issues he was facing in writing *An Approach to the Theory of Income Distribution* (1957). "Again a long lapse; perhaps there has been more excuse for it this time. I have been having another siege with the glands now quite high. . . . One must be constantly tenacious; good thing Nat is strong enough to bring this self-piteous vessel through the ordeals."[21]

He goes on to a mathematical issue: "You asked the following in your last letter. Is dZ/dN [divided by] $dV/dN = dZ/dV$ if Z and V are functions of N? Yes, always; look up method for finding derivative if curve given parametrically. Further you asked is dZ/dN [divided by] dV/dN (= dZ/dV) the same as $d(Z/V)/dN$? Definitely not." Hal then explains to Sidney that the derivative of a quotient is not the quotient of the derivatives.

This particular exchange, the question and answer, is important to the central argument of this chapter, and encapsulates Sidney's tragedy, both personal and professional. One of the things a high school or college student learns in the very first few weeks of a calculus course are the rules for taking derivatives of functions, and how to take the derivatives of particular functions. One memorizes, and is drilled on how to take the derivative of sums, products, and quotients of functions with the knowledge that the rule for quotients, and products, is not exactly intuitive although it is, once learned, "obvious." Sidney's question to Hal about whether the derivative of a quotient of two functions is the quotient of the derivatives of those functions tells the reader, Hal originally and us today, a great deal about Sidney Weintraub and mathematics. What kind of economics training would leave one, in the middle of the 1950s with a Ph.D. in economics, having taught graduate courses in mathematical economics at the University of Pennsylvania, to ask whether the derivative of a quotient of two functions is the quotient of the derivatives of those functions? There seems to be a serious problem in Sidney's mathematical understanding. The arguments that economist Sidney Weintraub wished to make are arguments in economics. They were not developed mathematically, but he was struggling to find an expression of them that was essentially mathematical. Such expression might present the material in such a way as to convince other economic theorists at least many of whom, by the middle of the 1950s, were accustomed to developing macro or micro theoretic arguments in mathematical terms. Yet in the mid-1950s, fifteen years after his Ph.D., a decade after he carried Osgood through basic training (and England, France, and Germany) while convincing himself that he was finally learning mathematics by himself, and several years after he had begun teaching graduate mathematical economics at Penn, Sidney could not take the derivative of a quotient of two functions. His mathematical skills were essentially those he had in the mid-1930s in college.

It is in this context that we must read Hal's fragmentary last letter. After the medical status report that generally begins these last letters, he wrote, "You no doubt have noticed that this is a long document. It is a first try at an idea suggested by some of my work on the physics project, and that has been used successfully in the study of gases. The possibility that it may be used with profit in economics has intrigued me, and so I send you an outline of fundamentals and definitions in the hope of your appraising the idea. Also, I have 'solved' a somewhat unrealistic problem by the method just to indicate the type of information that might result" (19 May 1954). Unfortunately, that "long document" has not survived. So we have no real idea of the kind of work that Hal was thinking of doing in economics, nor the kind of connection that Sidney and Hal might have had on work that was generated by Hal's ideas rather than Sidney's. What we can infer though is that even as he was in what would be his final illness, Hal was attempting to work with Sidney as a mathematician looking at problems in economics.

One wonders what Sidney made of that work. For myself, I believe that Sidney really did not, could not, understand such mathematical ideas. His was an outsider's appreciation that mathematics was coming to be important, and that there was a need for economists to write mathematically if they wished to have their ideas in economic theory taken seriously by the larger community of economic theorists. The frustration he expressed in his failure to get a hearing for his work in economic theory in the 1950s was not just a result of the newness, or the controversial nature, of those ideas that were eventually to be identified with Post Keynesian economics.

Paul Davidson has written (1985) that "as a refugee from the empirical research of the biological sciences with its emphasis on experimental design and statistical inference, I found Weintraub's realistic approach to economic analysis more relevant than the so-called 'scientific' empirical approach of some of my professors at Pennsylvania who tried to distill the values of economic parameters from time series data. . . . I was fortunate to study under Sidney during the period when he was developing his ideas for his *Approach* [*to the Theory of Income Distribution*] analysis; the lucidity of his arguments strongly affected my own choice of problems to be studied in future years" (533). But Sidney had to work out the final form of those ideas, which would be his most important legacy to the profession, alone.

> I loved Hal. When he was stung, I was hurt several-
> fold. When he succeeded, I was prouder than a parent.
> When he developed properly, I derived new courage.
> When he showed poor judgment, I grieved.
> I was happiest, vindicated maybe, elevated certainly,
> in inducing him to follow both a math, and a teaching
> career. He once wanted to be a journalist, a prize
> illustration of misjudging his talents. He could
> only write an unvarnished direct prose, that of the
> scientist rather than the man of letters. My judgment
> was clear here.
> He was discouraged at the lack of pecuniary rewards
> in teaching, and considered some applied fields, as

engineering. I dissuaded him for I thought he lacked
the salesmanship and the competitive urges and unitary
purposes that alone seem to insure industrial success.
As his illness developed, I was fortified in the
knowledge that my advice had been unerring.
With Hal goes my image, my confidante, my academic
confrere, my beloved mathematician, my brother in
the deepest sense. Much will be lost for me, so many
places are empty. Days despite my many loved ones are
open, long, and always incomplete.

It was not to be Hal who would have to help Sidney take the place he craved among the new generation of economic theorists. In 1964 I received an A.B. in mathematics from Swarthmore College, and in 1969 received a Ph.D. in Applied Mathematics from the University of Pennsylvania. Sidney and I were to "co-author/collaborate" on two papers: "An Inflation Unemployment Model" and "The Patinkin Full Employment Model: A Critique." We were never able, though, to establish a respect boundary that would permit his project to flourish. Not all fathers can be mentors to their sons. In any event, he never believed he had achieved appropriate recognition as an economic theorist. Looking back from today, and from the perspective of the preceding chapters, by the time he was ready to take his professional place, his intellectual time of nonmathematical economic theory had passed.

8   From Bleeding Hearts to Desiccated Robots

Putting it another way, we now have two streams of entrants into our profession—like the Missouri and Mississippi rivers joining near St. Louis. One branch—now the dominant one, in terms of the probability of attaining first-class citizenship—comes from mathematics, the physical sciences, and engineering, and its knowledge of history is lamentably small. The other branch, the second-class-citizen branch from which I come, issues from history and the other social studies; its weakness is on the mathematical side. What I'm hoping will be achieved some day, but have not furthered in my own career, is some sort of fusion and homogenization between the desiccated robots and the bleeding hearts, between pure technique in search of a problem and pure social consciousness in search of analysis.
—"A Conversation with Martin Bronfenbrenner"

Nothing is so difficult as not deceiving oneself.
—Ludwig Wittgenstein, *Culture and Value*

Looking around my own economics department, I see a number of colleagues of roughly my age, all of us trained in the 1960s. We were trained as undergraduates in mathematics at West Texas State, and at Cornell, in physics at Penn State, in engineering at Lehigh and at Georgia Tech, and mathematics at Swarthmore. How did we all end up in economics, and what has this meant for the discipline of economics?

Previous chapters have traced the complex interconnections between mathematics and economics over the first half of the twentieth century. However, the 1960s were different. If the 1950s witnessed the emergence of the community of mathematical economists, a slow acceptance of that community's practices, and an enlarged discursive space in which mathematicians and economists could engage one another's practices, the 1960s saw the results of that disciplinary evolution established in (U.S.) academic economics.

One needs to consider one demographic and one economic fact to fully appreciate the issues. First, as the "baby boomers" arrived as students in colleges and universities in the United States, beginning in the early 1960s, it was apparent that colleges and universities needed to expand in order to manage the education of that large new generation. Second, the postwar prosperity itself encouraged the Depression-era parents of those children to seek college degrees for their offspring. Higher education thus required more economists and more mathematicians. However, those who went into economics in the 1960s were different, in that their interests were differently constructed from those who had gone into economics in the 1930s. For while the latter were by and large impelled to study economics by the great economic social dislocations of the time, the Great Depression and the battle over economic ideologies—socialism, individualism—those who went into economics and were trained in Ph.D. programs in the 1960s were not similarly interested.

To understand this dissimilarity, let us go back to 1957, when I was just beginning high school. In October of that year the little beeps of the Soviet *Sputnik* satellite produced not only a political and military shock, but also a shock among American educators and thus for American popular culture. Of course, the explosion of the H-bomb had shown that we were not alone as a nuclear or thermonuclear power, but that at least could always be explained by the anti-Communists as the treason of spies. *Sputnik* was different. Why should the Soviets spy and then reconstruct, by imitation, our satellite launches? Our rockets had been failing. We (one of my uncles was a lead physicist in the Navy's Vanguard Program) could not get a satellite into orbit. We were chastened by the Communists' achievement. Their triumph of science and engineering led to a frenzied public agony about missile gaps, and science and engineering gaps in education. The result was a federally led revaluation of the importance in education of the field of science and mathematics. The very best students were now to be encouraged to study science and mathematics. Although I was not old enough to be drafted into the army, I had been drafted into the United States student army of the Cold War.

For teenagers in the late 1950s, if you were a good student, you were "tracked" into the science and math courses. The National Science Foundation, aggressively expanding into public education, had begun to pour money into school programs to enrich the science and math experiences of high school students. Beginning high school in 1957, just finding my way around the corridors of this new school by October, *Sputnik* and its effects were felt all around. National pride was at stake, and Communist Russians were not to be trusted. Mathematics, as the queen of the sciences, was regarded as the achievement for good students. If you were smart, you went into science and math, but if you were really smart, you went into mathematics. The geniuses went into mathematics. All who wished to be thought of as different, unusual, or special moved to mathematics. For the first time in many years, American high schools began looking at their best students as prizes, to be envied not pitied. "He (seldom she) is a real brain" became a term of at least modest approval for those years, replacing the "egghead" sneer directed at Adlai Stevenson. Of course it would be better to be able to throw a football 40 yards in the air, dunk a basketball, or get to first base with a cheerleader, but at least being a brain was something, finally. Nerds began to matter, if only a little bit.

With that kind of encouragement from the schools, the curriculum evolved rapidly. My suburban high school created courses in advanced physics, advanced chemistry, and advanced biology and put in extra years of mathematics (up to but not including calculus) as required parts of the college-preparatory curriculum. Putting aside for a time the pleasures of summer camps involving swimming, tennis, and hiking, I joined another student from my school between my junior and senior years as we were accepted to attend an NSF science camp at Northwestern University to learn about computers and such. And I entertained thoughts after that time of becoming a physicist or engineer or mathematician. In my house, Hal's memory was quite alive, and mathematics was more than encouraged for me. A family friend, Murray Gerstenhaber of the Mathematics Department of the University of Pennsylvania, provided me with materials (School Mathematics Study Group publications) from the committee on mathematics education of the American Mathematical Society, for that organization had taken on school curriculum revision to improve the mathematics standards in American high schools, junior high schools, and elementary schools. The New Math had been born in this movement.

In his autobiographical piece *Making It,* Norman Podhoretz (1967) opened by suggesting that, in our modern so-open culture in the United States, there was only one unmentionable subject. Questions about sex, religion, money all could be treated openly in a confessional literature, and everybody respects that there are even conventions for speaking appropriately about those subjects. Podhoretz then went on to argue that the only unmentionable subject in his 1960s world was "ambition." In the confessional world of the new millennium, of Jerry Springer and tabloid journalism, of presidential gropings as public jokes, ambition is no longer pornographic, with one rather curious exception. No scientist is permitted in an autobiographical piece to admit to personal ambition: Donna Haraway's (1997) modest witness (scientist) also witnesses the witness's modesty. Literary conventions for scientific autobiographies seem to require that all success, worldly success, result through superior intelligence (sometimes) and a series of accidents, local contingencies and good fortune. The larger autobiographical literature, since the early modern era's opening, has treated the life of a human being as a journey, and the path of the pilgrim making that journey is strewn with obstacles. The canonical autobiography then becomes a story of obstacles overcome on the path to enlightenment, or wisdom. Scientists, unlike others in this postmodern world, continue to write their autobiographical essays in this fashion. In their community nobody really goes out to backstab, to lie, or to cheat. One never seeks to redirect resources from others to make one's way in a scientific life, or so say the autobiographical literatures in science. Perhaps a scientist will admit that he did more to discredit a competitor's work than, in retrospect, he feels happy about. After all, we do have the record of Crick and Watson, in Watson's memoir *The Double Helix,* but the loud and irate discussions that followed that book's publication suggest that Watson really was contravening convention. And few believed him.

Ambition is tricky. Is it really very possible to distinguish between a desire to make one's way successfully in the world, as all others appear to be making their way successfully in the world, and something a bit untoward, something a bit more self-aggrandizing? Adam Smith, in his *The-*

ory of Moral Sentiments, speaks of a desire for emulation, a desire on the part of every human being to be looked up to by others for approval. Ambition is manifest in actions and attitudes designed, constructed, or otherwise developed in order to promote one's success. Even the TV tells us that it is right to "want to be like Mike."

But for those who can't dunk a basketball, who cannot run the 40-yard dash in 4.1 seconds on their way to tackling a frightened quarterback, for those who can't run a mile significantly faster than four minutes, for those particularly unendowed with physical beauty, or riches, jets, jewels or private islands, what scope might ambition take? For a scrawny moderately athletic Jewish kid growing up in the 1950s in America's northeast corridor, a kid with academic father, uncle, aunts, great uncles, etc., with rabbis and lawyers and doctors in the extended family in great profusion, ambition had to mean taking a place in that professional milieu. And if both one's father and mother believed, and argued, that doctors are quacks who kill people who fail to prepay for surgical operations, glorified butchers and technicians who earn too much money and act self-important, and if one's parents also believe that lawyers are either crooks who defend people for money not principle, or else who screw up the economy by passing laws about subjects of which they know little or nothing, the professional path appears clear. Approval from those parents, and internalization of parental values, required my becoming a professor.

Sidney Weintraub's embarrassment at having gone to college and graduate school for the Ph.D. at New York University led him both in his conversations and in his memoir (Weintraub 1989) to emphasize his graduate student days at LSE. His brother Hal, at Sidney's direction, was not to make the same mistakes, and so it was Sidney's pressure that led to Hal's decision to attend to graduate school in mathematics at Harvard, rather than engineering at MIT. For that professor's eldest son there were few permissible education options and thus, as I grew up wanting his approval, I internalized his perspective. By eighth grade (!), I had in my possession the catalogs of courses, the official university catalogs, from Harvard, Yale, MIT, and Swarthmore. Why those schools? Both Sidney and my teachers said they were the best, the most difficult to get into. Only the most accomplished students would be there, and my entering those schools would signal my honored specialness, and thereby his.

I was to leave for school in the fall of 1960, at age seventeen. Rejected at Yale (those were still the days of quotas for Jews in New Haven), I was wait-listed at Harvard (Sidney and Ken Galbraith spoke on the telephone and Galbraith offered his services to help move me into the accepted group). I never received the letter from my safety school, the University of Pennsylvania. Upset, I asked Sidney to find out the reason and he soon told me that he had called the admissions office and they said they never had received my application.[1]

Swarthmore it was to be, a hothouse of intellectual aggressors. Over the years as an indifferent student there, I majored in mathematics because, as so many there told me, "there is so much you can do with mathematics." This was not a period like that in which Kenneth Arrow, looking for an oc-cupation for someone with his mathematical abilities, sought out courses in statistics in order to become an actuary. It was not a time when Don Patinkin, loving mathematics, decided that there was no future in such a love and so began to study economics. No, by 1960–64, the post-*Sputnik* era, science and mathematics were the baskets in which the Cold War intellectual community had put its eggs. It was the patriotic duty of smart kids to become mathematicians and scientists. Ambition surely too led in that direction, for if the world were going to say "hooray for mathematics," it would a fortiori say "hooray for Roy."

At Swarthmore I eventually became a mathematics major, the decision overdetermined both personally and socially. Nevertheless it was there, finally studying mathematics, that the tensions in the American mathe-matics community filtered their way down to me. Whereas several decades earlier, a mathematics major would have done lots of other courses in physics and chemistry and engineering and the sciences, by the early 1960s mathematicians, under the Bourbaki influence, were no longer necessarily connected to the physical sciences themselves. A majority of my fellow mathematics majors were minoring in philosophy and physics, a popular combination instead of physics and chemistry, say. I more or less minored in philosophy and literature, avoiding science courses following my C- and D+ in introductory physics. And whereas this behavior might have cast grave doubts on my mathematical seriousness at the turn of the twentieth century, by 1960 it went unremarked upon: I even received some positive comments about being the only mathematics major graduating that year in the Humanities Division. Mathematics was beautiful, and theorems were described in aesthetic terms. Mathematical structure was to be explored and understood not in terms of applicability to other fields, but on its own terms with its own standards, standards where the highest compliments were of the form "What a lovely theorem," or "It's such an elegant proof."

I was not alone. My college graduating class had approximately 180 students, and 23 of us were mathematics majors. Thus, nearly 15 percent of my college class, my cohort who had begun high school with *Sputnik*'s beep, graduated as mathematics majors.[2] By 1960, all those who wished to go into a social science after graduation were encouraged to take more rather than less mathematics, but the mathematics we were learning was not directly applicable mathematics.

Training to Be a Mathematician

A few years ago in a conversation with the distinguished mathematician Phillip Griffith, now Director of the Institute for Advanced Study in Princeton, I mentioned my training and graduate study in mathematics in the mid-1960s, and he laughed and said, "You're a member of the lost generation." Mathematicians today, looking back to the period of the 1960s, shudder gently at much of what we then were expected to believe. For we were America's first fully Bourbakist generation of mathematics students, thoroughly inculcated with the ideals of Bourbaki mathematics, in love with structure, avoidant of applications. Bourbaki pedagogy, which moved from the general to the specific (the specific being examples), meant that one learned and understood about mathematical structures at the general level initially. One saw how groups, as a "mother structure," could motivate the discussion of modules over rings. Vector spaces over fields were at best a corollary, and finite dimensional vector spaces over the real numbers were simply left for problem 7, optional. Consequently if one had not had a thorough grounding in the concrete applications of mathematics, and understood at a practical and intuitive working level the origins of the ideas of groups from say transformation groups, the generalizations which defined the pedagogical moves remained all at an abstract level. As a first-year graduate student in mathematics at the University of Pennsylvania, I studied topology, algebra, and complex variables. The courses in topology and algebra were effectively conducted in French, using Bourbaki's volumes. An algebra course directed at multilinear mappings, and elements of category theory, competed for our attention with the nonstandard Bourbarki ultrafilter approach to topology.

In the period of time in which the computer was making its entrance, the mathematics department ignored computation. Computers were for electrical engineers, or maybe statisticians, and statisticians and engineers were intellectually lower class. One didn't want one's colleagues or one's students to slum. Those who washed out, failed out, or otherwise left the mathematics program were really being cast out of heaven, forced to live a more dreary and worldly existence. At the top, at the very top, the mathematician breathed the air of purity. After all, we lived on the top floor of the David Rittenhouse Laboratories building *above* the physicists.[3]

Accepted into the Pennsylvania Ph.D. program provisionally, I did passable work, but beyond the first year I did not do very well as a regular mathematics graduate student. Not having any understanding or appreciation, out of science literatures, where these structures that I was studying in mathematics had come from, I had no way of using or applying my intuition, for I had little or none, to figure out what was going on, to see connections. Perhaps some of this are my own idiosyncracies, but not having any grounding or cross-connecting in evidence, I found it nearly impossible to learn, at the mastery level, the Bourbaki mathematics that defined such a large portion of my graduate coursework. It was a struggle to get through preliminary examinations, a struggle that unfortunately was repeated a couple of times until I barely scraped through.

Just before the last time I took those examinations in 1967, the University of Pennsylvania, through Herbert Wilf of the Mathematics Department, received an NSF grant (again NSF!) to develop a Ph.D. Program Group in Applied Mathematics. That new group turned out to be ideal for me. The program as it was designed had as its rationale that Ph.D. students in mathematics would take, instead of three fields of specialization in mathematics, two fields in mathematics and one in an applied area. The students would then write a doctoral dissertation in that applied area either creating new mathematics for the application, or applying existing mathematics to a new application in the associated field. I remember, at that time, thinking that this was an ideal situation for me. I could continue to be a mathematician, and thus a worthy person, albeit an applied mathematician, and actually find something useful to do that I might even be able to do very well. At least in some other field my mathematics background would be a major status-creator, for so I had always been told.

I went to explore areas of applications, by checking out some of the individuals who were listed as the founding members of the applied mathematics group. I went first to talk to Duncan Luce, of Penn's Psychology Department, whom I knew had done a book on game theory with Howard Raiffa. We talked in his office for a while, but nothing came of it. Mustering my courage to face the inevitable, I casually mentioned my new occupational strategy to my father, who went across the hall and had a conversation with Lawrence Klein. In a day or two I received a short note from Klein suggesting that he had a couple of problems that he thought I might find stimulating were I to be interested in working in mathematical economics with him.

I recall walking around the streets of Philadelphia exhilarated, for several hours, thinking that I now had an escape from mathematical dimness, from a career that would not give me success at the level of my ambitions. Practically, the shift to economics promised such success, for the field was one that I was socialized to work in despite the fact that I had never studied economics.[4] Nevertheless, I did know something *about* economics even as I knew little economics, and I had some natural advantages in moving to economics. I recall thinking in a guilty fashion that I would have a much finer career, more consistent with my ambitions, in economics than I could in any other field. After all, it was my father's business.

I went to see Klein, and he outlined a couple of problems and I began doing a bit of reading. For the first problem he suggested involved my reexamining Kaldor's theory of the business cycle, and issues of nonlinearities. Klein believed that nonlinear differential equations needed to play a more significant role in modeling dynamic economic processes. Perhaps if I had stayed with that I could have written Hal Varian's paper on catastrophes and the business cycle before he did. However, the other problem that Klein posed involved general equilibrium theory, in particular an issue in the stability of the competitive equilibrium, namely how could one introduce random shocks into the tatonnement adjustment mechanisms to generalize the standard stability dynamics. He sent me to read the first four chapters of Henderson and Quandt's *Microeconomic Theory,* and then gave me Negishi's *Econometrica* survey article on stability theory to read. I was hooked, since I quickly saw that there was a nascent applied mathematical literature that bore directly on this problem. I could do that work. Another mathematician had thus come to economics.

The generation of economists that came of age intellectually between the late 1930s and the late 1950s began finding ways to talk across the boundary between mathematics and economics, although the boundary was real and the idea that economics as a discipline could appropriate ideas from mathematics was itself contested. However, the most important contingency in the history of that contest, World War II, divided mathematical economists from nonmathematical economists. By the late 1950s, the contest had been won by those whom Benjamin Ward and Terence Hutchison have called, inappropriately as we have seen in chapter 3, "Formalists." By the 1960s, there was no contest at all. Economics had changed its character, its language, its way of representing its own concerns. Economics by the 1960s had become a science of building, calibrating, tuning, testing, and utilizing models constructed out of mathematical and statistical-econometric-materials. And today, in the new millennium, economics remains that econometric community.

9 Body, Image, and Person

Historians of science are imposing order on segments of the past, often in a "narrowing" or "delimited" way. Something of this nature is inevitable when dealing with, to paraphrase William James, the great, blooming confusion of reality, past or present. The discourse of students of the history of science now admits more phenomena to the field, but without completely expunging older modes. Each of us can validly take a different slice of what we consider the subject. History of science is an eclectic field. Faced with any claim of an exclusive or superior path to historical insight, the prudent response is to walk away murmuring, "live and let live."
—N. Reingold, "The Peculiarities of the Americans, or Are There National Styles in the Sciences"

The preceding chapters have explored the interconnection of mathematics and economics in the twentieth century. The first story, beginning with Alfred Marshall and the Cambridge Mathematical Tripos, sets the stage for the new twentieth century in which the community of economists increasingly attended to mathematics, and ultimately remade itself as a mathematical science. I argue that reconstructing this set of changes requires attention to both the body and the image of mathematical knowledge, and the interconnected history of both the mathematics and economics communities. Using Peter Galison's (1997) language, mathematical economics became a trading site for the separate communities of mathematicians and economists; the exchanges they engaged in were sometimes characterized by semiotic impasses.[1] Indeed various chapters explored both the explicit and implicit links between those communities of mathematicians and economists, and what Pickering (1995) has termed the resistances and accommodations that both economists and mathematicians experienced in navigated the boundaries between the communities. Both directly and indirectly, I argue that the story of the intertwining of the twentieth century's mathematics and economics opens a dramatically informative window on both the structure and the development of modern economics.

Nevertheless, the question remains, "Is this the best way to tell the story of how the twentieth-century's economics was remade?" If it is not the best way, what other ways are available to us? What are the historiographic alternatives?

There is a long disciplinary tradition of writing histories of economic ideas in which continuity and progress are the main protagonists. The economic theory community, well versed in mathematical ideas and with some knowledge of the internal history of mathematics, anticipates a story in which a perhaps intellectually courageous economist brings forward a new set of ideas. Theorists look for stories of progress, and Nobel Memorial Prizes validate such a belief as reasonable and natural.

Other economists may be less enthusiastic about reading a narrative that is constructed as a claim for progress. Many individuals, ambivalent about the mathematization of economic theory, are likewise ambivalent about a story in which mathematization is appraised as progressive, heroic, and knowledge-increasing.

My own previous historical writing can suggest some of the issues that these questions raise, for the present volume has continued my investigations into the development of economics in the twentieth century. The first of these historical studies appeared as an appendix in my doctoral dissertation (Weintraub 1969), and typifies that which the historian Ted Porter (1992, 235) has lamented:

> Unfortunately, many historians of economics are so completely socialized as economists, and so little as historians, that the genre of historical study is not fully distinct from that of the review essay. The review essay surveys a field and assigns credit, usually on the assumption that knowledge is steadily progressing. Far too much history of economics, still, aims to extend the review back twenty or fifty years by presenting the ideas of the economist on some modern question. The precursor, long dismissed as a category mistake in history of sci-

ence, is still alive and well in economics, and this is almost inevitable so long as history of economics is written to meet the standards and presuppositions of ahistorical economists.

Put another way, the narratives economists and historians separately construct about the history of economics are often incommensurable. Most of us economists are simply in the business of doing economics. We are not historians nor are we particularly concerned with the evolution of ideas.

My own historical writing has been "a long struggle of escape" (Keynes 1936, xxiii) from the scientist's understanding of the nature and role of the history of science, to the historian's.[2] In addition, my own journey recapitulates the evolving historical concerns of my new colleagues in the history of economics subdisciplines.

Viewing Science

For a working economic scientist like me, most of the histories of economic science I had read were shaped by my interest in the past of my own subject. However, if an extended survey was insufficient to "tell the story," it was likely the case that my economist's framework for a potential narrative line was inadequate. Where was I to find a more acceptable framework? Where were structures or theories about the development of economic science to be found? Who was interested in theorizing science anyway?

As a Swarthmore college student, I had had a course and a seminar with the philosopher of science Lawrence Sklar and I had learned about covering law theories of historical explanation. I had also read enough to realize that philosophers of science were professionally engaged in examining the history of science as material for testing and refining their notions of what constituted good and bad science. My increasing interest in history thus was shaped by my having come to the history of science, and thus the history of economics, through philosophy of science, or what in economics is called "methodology."

In his paper "History of Science and its Rational Reconstructions," Imre Lakatos (1971) argued that every particular philosophy of science, that is, every system that develops a normative reconstruction of science and the development of scientific knowledge, carries with it an associated historiography of science. It is not the case that history of science provides case studies for philosophers to test alternative conceptions of how science operates, but rather that each particular conception of how science operates constrains the narratives that can be constructed in the history of science: "Each internal historiography has its characteristic victorious paradigms" (ibid. 104).

Consequently for Lakatos, "The inductivist historian recognizes only two sorts of genuine scientific discoveries: hard factual propositions and inductive generalizations. These and only these constitute the backbone of his internal history. When writing history, he looks out for them [even though] finding them is quite a problem" (ibid.).

More generally, each method for appraising scientific work attempts to distinguish successes from failures in science, and each defends the "right" method as the one that produces successes. Consequently, each alternative methodology for economics would appear to have implicit winners and losers in economic work: the winning economic ideas, those that emerged from the community's work, are exemplars of the right methodology. *In approaching history from philosophy, the history of economics becomes a history of winners and losers with respect to particular theories of good and bad economic science.*

To write history, I (and others) simply had to get the right methodology for understanding economics, for once we knew what good economics was, what parts of economics could be positively appraised, we could narrate how they had come to be that way.

Inductivist Heroes

This would have been an easier task a hundred years ago. At the turn of the twentieth century, science, a fortiori economic science, was generally understood to be inductivist. Many educated individuals had read Karl Pearson's *The Grammar of Science* (1911), published in various editions between 1892 and the first decades of the twentieth century, and supposed it to be a coherent picture of exactly how science proceeded. Pearson summed up his discussion of the method of science by arguing that "the scientific method is marked by the following features: (a) careful and accurate classification of facts and observation of their correlation and sequence; (b) the discovery of scientific laws by aid of the creative imagination; (c) self-criticism in the final touchstone of equal validity for all normally constituted minds"(37). Pearson, a phenomenologist, argued that the external world provided sense impressions that the human being interpreted through brain activity. Using the metaphor of the brain as a central telephone exchange, Pearson's vision had the scientists operating as interpreters of the messages from nature, as they went about classifying and reconstructing data. The facts of science, and thus the facts of economics, "excite the mind to the formation of constructs and conceptions, and these again, by association and generalization, furnish us with the whole range of material to which the scientific method applies" (ibid., 74).

If we think that science proceeds through the accumulation of instances and the construction of theories by building on data, and rationalizing data to generate new ideas that themselves can be confronted with data, we can construct a quite coherent narrative about the progress of economic thought in the twentieth century, and our hall of heroes quickly fills with some statuary. The historiography would certainly feature the work of the National Bureau of Economic Research, with Wesley Clair Mitchell and Arthur Burns and later Milton Friedman, Anna Schwartz, and others. It would feature as well work on business cycles connected to the NBER project, and sponsored by the Rockefeller Foundation in institutes in Europe in the interwar years (Rotterdam, Vienna, Kiel,to name a few), placing individuals like Jan Tinbergen, and perhaps Ragnar Frisch, in positions of prominence. Our stories of scientific success would highlight the economic ideas of Wassily Leontief, whose careful classification system of input-output tables allowed an ever finer representation of the structure of particular economies, and the usefulness of those representations in managing the command economies of counties in the former Soviet bloc. The inductivist pantheon would include Simon Kuznets, and James Meade and Colin Clark, whose development of the ideas of national income accounting gave prominence to the collection and classification of the facts of the domestic economy, facts that could be arrayed and understood in a Baconian fashion to allow theorizing to proceed. The work of Edward Denison and Moses Abramovitz would likewise appear with prominence in the stories of the successful work in economics. Related stories would address the activities of the Cowles Commission in the United States, and the League of Nations statistical work, and the activities of the Bureau of Labor Statistics in the United States, and the statistical offices of the ILO, OECD, etc., for econometrics grew out of such activity. Trygve Haavelmo, Lawrence R. Klein, Herman Wold, Abraham Wald, Tjalling Koopmans, and so many others figure large in the tale.

If inductivism is the proper methodology for economics, the proper history is a story of increased content, and increased facility by economists and their allies (demographers, statisticians, etc.) in providing accounts of the world that are useful for description and control.

Critical Rationalism

However, writing histories of economics late of the twentieth century, from a methodological perspective, forced attention away from inductivism, and toward the writings of the philosopher Karl Popper (1959). Since the late 1930s when Popper initially was packaged for economists by Terence Hutchison (1938), his ideas have been thought by some economists to provide a compelling normative statement of how economics should be done, and why economics done in that way could be a real science like physics. Popper's argument was that science proceeds by a series of conjectures and refutations, by which bold hypotheses are ruthlessly subject to attempts at falsification. A real science holds all propositions and theories to be provisional, while serious scientists attempt to refute particular conjectures or theories. Science progresses by the weeding out of error and this self-correcting process is what is meant by scientific progress. To write the history of science, in this view, we need to look at exemplars of good science, at instances where knowledge was gained by the eradication of error. However, if for physics the paradigm of such good science was the Michaelson-Morley experiment,[3] which failed to find evidence of the luminiferous ether, it is very difficult to see evidence of progress in economics. The falsification of economic theory by empirical/statistical evidence is virtually unknown to economists: as the eminent historian of economics Mark Blaug (1980) has written, economists practice "innocuous falsificationism." For a historian wedded to the Popperian view, twentieth-century economic thought is a melange of prescientific musings about social problems wrapped in the language of science, without any real science in evidence, and writing histories of economics is akin to writing histories of phrenology.

Undiscouraged yet eager to find progress in economics, some economists pursued a more tolerant variant of critical rationalism that developed from the writings of Popper's Ph.D. student, Imre Lakatos (1970).[4] For Lakatos, science is done within what he called a "scientific research program"; the program consists of a set of ("hard core") propositions held to be true and irrefutable by those working in the program, associated rules for constructing theories based on those central premises ("the positive heuristic"), and rules ("the negative heuristic") for excluding, as uninteresting or irrelevant, material outside the purview of the program. Scientific analysis is carried out in the "protective belts" of the scientific research program, which consist of theories developed from the heuristics. Progress occurs as the scope of the program is extended to handle previously anomalous cases that are explained by the theory or theories in the belts.

This framework has appealed to methodologists, and I can speak with some certainty here because I was one of them. We told the story of twentieth-century economics as the rise and fall, or the progress or degeneration, of various scientific research programs in economics (Latsis 1976; Weintraub 1985; Blaug and DeMarchi 1991). Among programs, the neoclassical research program would be likely to be the most successful since it has had the largest number of economists working within it. Its various hard-core propositions of optimization subject to constraint, appropriate assumptions on knowledge, and rules for constructing models based on such principles (while avoiding building theories based on irrational activity, or changing tastes, and so on—the negative heuristic), have been extended and deepened over the course of the twentieth century so that the theories associated with the neoclassical research program have themselves stabilized various knowledge claims in economics. If there is a mainstream, this is it. There are alternative programs, partially overlapping in some cases with the neoclassical research program. One might think of the Keynesian program as a particularly interesting one that developed in the 1930s and was successful and progressive through the 1970s when its predictive failures and theoretical difficulties, brought out by its confrontation with simultaneous unemployment and inflation, led to its relative degeneration with respect to the alternative neoclassical program in its New Classical form.

In terms of alternative programmatic discussions, and narratives of the waxing and waning of particular varieties of economic thought in the twentieth century, we can see the continuous degeneration of what could be termed the Marxian research program as its anomalies could not be incorporated without ad hoc changes in the hard core of the program.[5]

Nevertheless, constraining historical narratives to identify characteristics of a Lakatosian program imposes a rational, not a historical, reconstruction on all the materials, and thereby constricts the narrative. To tell the story of the development of modern labor economics, the rational programmatic history focuses on particular features that may or may not be historically explanatory. For instance, the history might require a detailed sensitivity to the nuances of data collection and construction, but those features of data analysis are hardly touched on by a rational reconstruction: for such, data simply "are." The Lakatosian version of the "confrontation of theory by evidence" is as historically unhelpful as the Popperian story. Nevertheless, a Lakatosian framework produces histories of progress and degeneration, the rise and fall of congeries of ideas and theories and hypotheses and evidence and training centers, and provides thus at least a sense of the vitality of economic science. For that reason perhaps it still provides economists with a sense of a heroic past, though the protagonists are the programs, not the people.

Revolutions

In the 1960s, economists were captivated by Thomas Kuhn's persuasive account of how science proceeds in his *The Structure of Scientific Revolutions* (1962). Kuhn argued that in any particular time and place, science operates with an established vision of the relevant disciplinary world: the intellectual framework of the science, and an understanding of the problems that are open and unresolved within that framework, is the paradigm of the science. Scientists working within the paradigm, engaged in what he termed normal science, are solving the natural puzzles and problems that arise in the course of doing the work. On some occasions though, the established consensus begins to break down. Perhaps an anomaly is obvious, and awkward, so that accommodating it leads to incoherence within the established paradigm. Perhaps certain experimental results, or phenomena, or analytic issues, lead a number of individuals to see the scientific work differently. Such periods are what Kuhn called "crises," and he identified them as harbingers of revolutionary episodes, for they change the fundamental culture of a scientific field: that which was understood is no longer understood in the same way as people literally see and think differently. For Kuhn, there are infrequent episodes in which scientific discourse changes in an irrevocable way: such ruptures he termed scientific revolutions. Moreover, and it is a more controversial claim, for Kuhn one of the features of revolutions is the fundamental incommensurability between the visions instantiated in the prerevolutionary paradigm and the postrevolutionary paradigm. Individuals literally do not understand the subject in the same way as they did before the revolution. The problem-syllogism for economists became: "Science has revolutions. Economics is a science. Therefore, economics has revolutions." But has economics had revolutions in Kuhn's sense?[6]

A number of historians of economics, and economists using the language freely, have answered "yes." For them, economic thought in the twentieth century is a narrative of discontinuity: they speak of the marginalist or neoclassical revolution, the Keynesian revolution, the monopolistic competition revolution, the Sraffian revolution, the rational expectations revolution, the game theory revolution, the econometric revolution (Black et al. 1973). From talk of the neoclassical or marginal revolution, to talk of the Keynesian revolution or the econometric revolution, there is a sense of a break with the past in a comprehensive fashion. For example, it was argued that discussions of unemployment prior to Keynesian macroeconomics became literally incoherent from a Keynesian framework while from a rational expectations perspective, Keynesian involuntary unemployment literally makes no sense. Incommensurability of conceptual frameworks across paradigms explains why Robert Lucas and James Tobin do not argue with one another in any productive fashion. It can explain why John von Neumann and Paul Samuelson each thought the other quite foolish concerning game theory, but why Samuelson and Tobin can argue productively about stimulating economic growth. From the Kuhnian perspective, we have not in fact two competing theories that can be appraised, one against the other, based on tests or the evidence per critical rationalism. Instead, we have two alternative visions of the workings of the economy itself, with alternative vocabularies, rules for linking concepts, and understandings concerning the nature and significance of the interconnections (Dow 1985).

The language of normal science and revolutionary episodes induces a heroic historiography. In economics, as in other disciplines, such a romantic vision recommends itself to the practitioners. Many economists quite favor the idea that the history of the twentieth century is a set of chapters recounting how individual economists, with courage and tenacity, changed the nature of practice, and so economists favor histories of economics that look at the discontinuities in economic thought, the breaks, and attend to the features of the intellectual and cultural landscape which lead up to, and lead away from, the revolution.[7]

My own first serious attempt at writing for historians of economics reflected this Kuhnian move. Early in the 1970s, still attempting to establish a respect boundary with my father, I wrote on a topic close to his own interests of "Keynes versus Keynesians." In my "Uncertainty and the Keynesian Revolution" (1975) I argued that a feature of Keynes's "real" revolutionary message had been lost in the postrevolutionary fervor. However, that paper failed to find a place in historical arguments of others even as it lived on for a while in Post Keynesian circles. Reflecting on this over the years, I have come to see why many historians of science abhor such histories, and to understand why the historian's perspective is rather uncongenial to most economists.[8]

> The aim of historical scholarship is to demonstrate that science is a genuine historical process shaped by and shaping social and political agendas. *The practicing scientist has no privileged access to this history.* . . . The fact that such an exercise is deemed to be subversive by scientists, underscores *the essential tension* between the two professions. . . . Scientists' history is often reduced to a collection of anecdotes, or, as for instance in historical introductions to textbooks or also in personal accounts, presents a rational reconstruction of the development of scientific theories. In these accounts history proceeds by theoretical breakthroughs attributed to scientists of particular brilliance and in-

sight. These histories often serve disciplinary needs like constructing a research tradition or legitimizing a new research field. (de Chadarevian 1997, 61; emphasis added)

For a historian, that "essential tension" produces not a merely a problem of narrative. For example, the historian regards most economists' stories of the Keynesian revolution as a historical wasteland littered with legitimizing accounts of the nature of the specific break that made the difference (Uncertainty? Effective demand? Liquidity preference? Futures markets? Money? Involuntary unemployment? Wage rigidity?). From the historian's perspective, only recently have we had productive work in this area by scholars like Peter Clarke (1998).

OTSOG-ery

Related to the revolution-induced histories, we may also identify another less well-organized historiography, one which has been termed "On The Shoulders Of Giants" (OTSOG), recalling Newton's claim that he saw so far only because he stood on the shoulders of giants.[9] Such heroic visions of science call us to heed times when giants walked the earth (Chicago in the 1930s, Harvard in the late 1930s, Cambridge in the years of high theory, LSE in the 1930s, MIT in the 1960s, Minnesota in the 1970s, etc.). A history of economics reflecting this perspective then is a chronicle of the greats, and accounts of their interactions and contributions. For what it is worth, this seems to be the accepted historiography of the committee that awards the Nobel Memorial Prize in Economics. It certainly directed my own past writing on general equilibrium and dynamics (1985, 1991) as I provided accounts, organized by chapters or sections, of the work of Paul Samuelson, Kenneth Arrow, Gerard Debreu, John von Neumann, Lionel McKenzie, and others.

Science Studies

From all of these methodological perspectives—inductivist, critical rationalist, normal science/revolutionary science, OTSOG-ery—the historian of economics works with a metanarrative of progress. As the philosophy of science privileges scientific knowledge over mere belief, scientific demonstration over mere argument, science over magic, the roots of that privilege are located in the advancements of science itself. Through one mechanism or another—its error-correcting features, its democratic openness, its value-neutrality—science and progress are linked. As error is weeded out, truths are uncovered, and knowledge claims are stabilized through the application of particular and specific methods.

Most economists are comfortable with such narratives of progress. To be socialized as an economist in school is to learn the current tools, techniques, methods, and appurtenances of the discipline and thus to have a great deal invested in the rectitude of current ideas. Today is better than yesterday, for after all we are learning (investing our scarce human capital in) today's theories. Most working scientists (economists) are Whig historians who believe that the best of the discipline's knowledge and practices are contained in the current material of the discipline. Therefore, narratives constructed to lead up to the present in a progressive fashion seem in fact to be truly how it is. To argue otherwise would appear to be either a quaint antiquarianism, concerned for old ideas for their own sake (whatever that means), or else a misguided critical attempt to attack current ideas based on historical analysis of their origins (Keynesianism is wrong because "Bastard Keynesians" misinterpreted Keynes's chapter 17).

However, ask how can one write the history of England or America in the twentieth century, or the history of French diplomacy in the twentieth century, or the history of Soviet aircraft engines in the twentieth century? Is not the issue how one writes any history? Once the problem is framed in this way, it becomes clear that a metanarrative of progress is but one historiographic alternative.

My own path to an alternative was through an emerging literature critical of philosophers' normative accounts of science itself. The new writing developed a perspective based not on asking of science how it should be done, but rather how it was and is done. The naturalistic turn—thinking about science by actually looking at how science is done and what scientists do—is best represented by a group of sociologists of science, later joined by philosophers and historians, and others, under a banner called "Science Studies" or "Science and Technology Studies," or "STS" for short. This approach to thinking about science, and thus thinking about economics, looks very specifically at practice, at the real engagement of individuals with human and nonhuman materials. From discussions of human and material agencies, the resistance that the materials present to human agency, the mutual stabilization of thought and practice, of theory and evidence, or data and experiment, of belief and knowledge arises a view of science as a craft, an activity in which real people do real things. Lost is the grand vision of revolutionary episodes, theories confronting data, and progress associated with greater and better knowledge about the external world. What replaces such stories are local narratives of laboratory life, of technological innovation, of ideas transformed by argument.

For example, if we choose not to tell the story of twentieth-century economics with a narrative of scientific progress, we could employ the vocabulary of Bruno Latour's (1987, 1988) actors and networks. This particular STS framework might begin with economists doing economics, employing arguments, and working with representations of economic behaviors that take shape, and gain epistemic power, through their instantiation in networks of ideas (e.g., National Income), calculations (e.g., the National Income Deflator), representations (e.g., the open economy), institutions (e.g., the Council of Economic Advisers), etc.[10] Beliefs become knowledge in the relevant communities as the networks are extended, and more agents and more networks support the beliefs. Scientific knowledge, for Latour, is extremely robust as it is extended through so many material and nonmaterial actors. Latour's vision depicts scientists attempting to extend their networks, to interlink them with others, to overcome obstacles and to win tests of epistemic power called trials of strength by overcoming objections of other scientists, by obtaining better results in well-understood contests called predictions.

This kind of argument lay behind my own reconstruction of the modern history of economic dynamics (Weintraub 1991). As I then (mistakenly) believed myself freed from the tyranny that a narrative of scientific progress coerced me to write, I argued that there were better ways to approach the history of economics in the twentieth century. I could, for example, examine interconnections between economics and other discourses of other disciplines without expressing an opinion, one that would necessarily shape the story of course, about whether such interconnections were "good for economics" or "bad for economics." Histories of the interconnection of mathematics and economics simply provide a different framework for talking about the changes in economic ideas, and the forms in which those ideas are expressed.

Attending to such notions produces different histories of economics. The histories allow us to ask interesting and complex—"thick" (Geertz 1988)—questions: if there is no presupposition that there is one and only one right way for economic analysis to proceed, one and only one way in which economics can modify itself, our histories can reconstruct economists engaging in controversy and ending those controversies. From such a perspective, economics in the twentieth century becomes a human activity in which many individuals are engaged in a complex, locally situated, and contingent conversation, where the rules for community membership are fluid and conventions of discourse are communally well-understood.

What of course I failed to realize was that there could be no escape from "frameworks." There is no view from nowhere, no platform on which I, the historian, can stand apart and aloof from the materials on which I work.[11] Science studies provided me with the right perspective, which is to say a perspective that was useful for my purposes. That perspective is not, and cannot be, privileged. Yet giving up the search for the best way to write history has allowed me to explore good ways to reconstruct the past and thus to establish useful truths about the history of economics.

Socialization

Too often histories of economics engage the charming conceit that economic ideas are autonomous free-floating ethereal objects, which pass from one disembodied mind to another quite unmediated, though they are occasionally transformed by other products of pure thought. The evidence our professional lives provides a reality quite different. Real people (like you and me) have beliefs, those beliefs are what we take to be ideas, and these ideas are transformed, reconfigured, and reinterpreted in cascades of representation and re-representation in intentional (and sometimes unintentional) discourse communities. But what do we know about that community of economists who somehow are responsible for having, and transforming, economic ideas? As infants are not born speaking a language of supply shocks and heteroskadasticity, the process by which individuals become economists conditions and shapes the practices, including the speech practices, of those who identify themselves as economists.

How then can we talk about economic ideas without having an understanding of how economists are trained in the twentieth century? For many economists, and for some historians of economics, these questions smack of personalia. George Stigler, for example, bemoaned the idea that biographical studies of economists had any place in the history of economics, arguing that Marshall's laundry lists are not data for the historian of economics. Stigler was wrong. The contingencies of time, and place, and experience are not independent of the ideas that are expressed in time, in place, and in experience. People hold economic beliefs, and beliefs are shaped by personal and social experience. Biographical studies, and sociological studies, of the education of economists are too infrequently done. The differential nature of national economics education and training is relevant to the ideas of economists writing in different languages in different countries (Coats 1993, 1996). The presence of Ph.D. training in the United States, and its general absence in the United Kingdom before the 1970s, is one kind of relevant context for understanding the kind of work that the British economists themselves did. How economists were socialized to be economists in France, and Italy, and Sweden, and Australia, and Japan is dependent on the kinds of ideas that those economists found congenial, and the kind of work that they would do later as professional economists. A sympathetic and systematic understanding of twentieth-century economic thought must account for variations in those thoughts at least in some measure by providing an account of the developmental context in which those communities thought those thoughts. The particular techniques, tools, and habits of mind of economists do not appear full-grown from the head of Zeus: rather they emerge imperfectly from the educational practices that inculcate certain habits of mind and techniques of craft.

We have few examples of this kind of historical writing. Most of the biographies of economists, written by economists, are silent on such matters. Autobiographical accounts, solicited by economists and publishers, conform to the OTSOG formula, with the modesty of the economist shining through the theories that established his fame. My attempt, in chapters 7 and 8, to write that kind of account, is one of only a few such.[12]

Taking the History of Economics Seriously

Although the history I have constructed is not a methodologist's exemplar, nor is it engaged with ideas of progress, neither is it a particularly theorized account employing the most recent artillery in the Science Studies arsenal. One of the advantages in having constructed histories in the past in accord with one or another framework for writing about science is that I have less at risk in freeing myself from the imperatives of "the new best way to write history." Indeed, I no longer believe that there is such a best way (which does not mean that there are not good and useful ways) to write history.

There is no definitive book called "The History of Economics in the Twentieth Century." I believe that an attempt to construct such a history will bear little resemblance to Schumpeter's magnum opus suitably updated, nor is it likely to resemble older textbooks with chapters on Physiocratic Thought, Mercantilism, Adam Smith, etc. As each generation writes histories consistent with narratives of what our economics discipline is "supposed" to be doing, the histories themselves will shift in their perspectives. The twentieth century differed from the nineteenth, and just as certainly twentieth-century economics differed from nineteenth-century economics. We have seen economics become a social science, taking its place among established scientific disciplines. It even awards Nobel Prizes to its "stars." Public discourse greatly respects economists, attending to economic advice that is seriously given, and often taken. We have seen economists trained as professionals, and educated in a worldwide network of ideas and understandings. New techniques, new ideas, and new approaches to being an economist have energized the field of study, and practice of economics. The history of economics then must recognize both the diversity of perspective and complex richness of this human practice called "doing economics."

In the preceding chapters, I tried to practice what I have been here preaching. The history that I constructed explored some of the connections between economics and mathematics over the twentieth century. In exploiting the body-image distinction with respect to mathematical knowledge, and introducing the author as actor and site, it has proffered one among many possible alternative readings of the past century's economics, one among many possible alternative stories of the history of economics. It cannot be the best story, for there can be no such story. Nonetheless, it is my hope that it can illuminate, entertain, teach, and thus lead to a change in our collective beliefs about our past, which after all is what we seek in any history.

Notes

Prologue

1   It is not as if the historiography of mathematics itself is so settled that it can be simply applied to my own project. In their introduction to a set of papers exploring the history and philosophy of mathematics, William Aspray and Philip Kitcher (1988) presented an overview of approaches to the history of mathematics, suggesting the kinds of shifts in historiography associated with, but lagging behind, changes in the ways the history of science were written in the decades following the founding of that discipline by George Sarton. They noted that

> With the advent of a professional history of science, a new and more sophisticated historiography has arisen and is being put into practice in the history of mathematics. This historiography measures events of the past against the standards of their time, not against the mathematical practices of today. The focus is on understanding the thought of the period, independent of whether it is right or wrong by today's account. The historiography is more philosophically sensitive in its understanding of the nature of mathematical truth and rigor, for it recognizes that these concepts have not remained invariant over time. This new historiography requires an investigation of a richer body of published and unpublished sources. It does not focus so exclusively on the great mathematicians of an era, but considers the work produced the journeymen of mathematics and related scientific disciplines. It also investigates the social roots: the research programs of institutions and nations; the impact of mathematical patronage; professionalization through societies, journals, education, and employment; and how these and other social factors shape the form and content of mathematical ideas. (24–25)

2   Let me introduce a caveat: most of the literature on the development of mathematical economic theory, indeed most of what is canonized in modern economic theory itself, is a literature in English. Yet the world of mathematics involves at least significant

French, German, and Russian mathematical communities, and likely Polish, Dutch, Italian, Hungarian, and Swedish as well. The peculiar differences between the experience of English economists with mathematics and those of "Continental" economists are quite important to keep in mind, for discussions about the development of, the nature of, and the possibilities for the mathematization of economics take on very different meanings when they move from England to France to Germany to Italy. While it may be comforting to speak about the one world of mathematics, and the unity of scientific discourse, such is hardly ever the case.

1   Burn the Mathematics (Tripos)

1   For example, the title of a well-known history of modern mathematics is *Mathematics: The Loss of Certainty* (Kline 1980).

2   Following Richards (1988) I shall capitalize "Tripos" except in direct quotations where the author uses the lower case.

3   Carpenter took religious orders, and had a parish ministry, but left following a breakdown brought on by his reading of Walt Whitman's poetry. "He was a rebel who revolted against everything Victorian: their narrow view of spirituality, their obsession with money and capital, their narrow views of human potential and worth. Edward Carpenter was involved in all the main progressive movements of the late 19th and early 20th centuries. He was a Fabian Socialist and a friend of William Morris, a Vegetarian, an anarchist and communist, a spiritual adept and sexual revolutionary" (Grieve 1996).

4   Marshall, in his 1871 "Essay on Wages" (Whitaker 1975, I, 186), wrote that "the fact that the farmers fix the price tends to make the average price lower than it would otherwise be: just as you make the average volume of a hollow partially elastic Indian rubber ball with a hole in it smaller than it otherwise would be by continually pressing it in." I am indebted to Simon Cook for calling my attention to this reference.

5   Whewell (1845, 40–41) argued, in support of what were to be the 1848 Tripos reforms, that "in the one case, that of geometrical reasoning, we tread the ground ourselves, at every step feeling ourselves firm, and directing our steps to the end aimed at. In the other case, that of analytical calculation, we are carried along as in a railway carriage, entering it at one station, and coming out of it at another, without having any choice in our progress in the intermediate space. . . . It is plain that the latter is not a mode of exercising our locomotive powers. . . . It may be the best way for men of business to travel, but it cannot fitly be made part of the gymnastics of education." I am grateful to Simon Cook for providing me with this reference.

6   Marshall of course was not alone in having misperceived the changes in mathematics: educators were still toeing the old party line at century's end: "Mathematics in its pure form, as arithmetic, algebra, geometry, and the applications of the analytic method, as

well as mathematics applied to matter and force, or statics and dynamics, furnishes the peculiar study that gives to us, whether as children or as men, the command of nature in this its quantitative aspect; mathematics furnishes the instrument, the tool of thought, which we wield in this realm." (Harris 1898, 325, as quoted in Moritz 1914, 62).

7   Of course, Marshall was aware of these new ideas in mathematics since he was close to W. K. Clifford who introduced such ideas about non-Euclidean geometry to England. For instance, see Marshall's account of his meeting, in the United States, with Ralph Waldo Emerson (Whitaker 1996, vol. 1, 62). Simon Cook observed that "although Marshall may have been abreast of Clifford's work, he did not really embrace it. . . . He saw that there were philosophical consequences—his fourth paper to the Grote Club the 1869 'The Duty of the Logician'—shows that he saw non-Euclidean geometry as supporting Spencer's evolutionary account of mind over Kant's notion of the *a priori* as outside of time" (Cook 2000).

8   The British Association (for Science) did not really recognize Economics as a separate discipline, lumping it together with Statistics and other social sciences in a catchall Section F, behind other sections for Physics, etc. It was not until Marshall organized the Royal Economic Society that British economics would have a distinct organizational identity.

9   Mirowski goes on to argue how enmeshed Edgeworth became in the trappings of the Oxford professorship and editorship of the Royal Economic Society's *Economic Journal*. He argues that Edgeworth's increasing isolation was a result thus not only of the exterior intellectual menu shifting, but Edgeworth's having to contend continuously with Marshall and Marshall's school.

10   The version that I shall examine is the translation by Ludovic Zoretti into French for *La Revue du Mois* of 10 January 1906, where the paper was called "Le Mathématiques dans les Sciences Biologiques et Sociales" (Volterra 1906b). In what follows, I shall use an English translation of this French version prepared for me by Caroline Benforado in December 1993.

11   These functions play the role of utility functions, whose level surfaces are the indifference curves, or iso-utility curves, of neoclassical microeconomics.

12   This characterization of mathematics originated with Gauss. See Moritz 1914, 271.

13   There is ready evidence that Marshall was aware, at some level, of this changed image of mathematics. In a letter written to the American economist Walker, he stated: "I am very much impressed by the enormous advantage you have over a man like myself, for example, in being a mathematician and a physicist. I shall have to qualify that remark. The advantage I have in mind chiefly comes from your being a physicist. I don't so much envy the mathematician, tho' I can readily see that he has a great power of illustrating mathematical truths, and of expressing them in terms at once compact in themselves and familiar and welcome to many minds. But the physicist (who might, I

suppose, conceivably be not even a good mathematician) has a truly enormous advantage in studying the phenomena of industrial society, in watching the propagation of economic shocks, in tracing the lines of fracture from commercial or financial disasters" (letter to Marshall from Francis Amasa Walker, 16 October 1890, as quoted in Whitaker 1996).

2   The Marginalization of Griffith C. Evans

1   The title of this section is taken from the lovely paper by Judith Goodstein (1984).

2   There appears to be no full-length biography of Volterra, and certainly none in English. There are, however, several biographical essays in English, the first of which to appear was written by Sir Edmund Whittaker in 1941 as an obituary notice of Volterra as a fellow of the Royal Society. This essay, which contains a virtually complete bibliography for Volterra, was itself reprinted by Dover in Volterra's book on *The Theory of Functionals* (Whittaker 1959), and was the basis of the "Volterra" entry in *The New Palgrave* (Gandolfo 1987). One other source of details on his life can be found in the note by E. Volterra in the 1976 *Dictionary of Scientific Biography:* that piece has an exceptional guide to the secondary literature on Volterra and his role in Italian science and mathematics (Volterra 1976). Most of my own thinking about Volterra has been influenced by Giorgio Israel, who has written extensively on his life and work. The references to Israel's work can be found in the bibliography.

3   See the very good compilation contained in the five volumes of Volterra's *Opere Matematiche* (Volterra 1957).

4   Again, let me acknowledge my debt here to Giorgio Israel whose own leadership role in interpreting the history of Italian mathematics has shaped the views I set out here.

5   I thank Giorgio Israel for several conversations on this matter, and for insistence on this distinction. His own paper "'Rigor' and 'Axiomatics' in Modern Mathematics" (1981) shapes the discussion.

6   Volterra himself, writing this in 1901 when he was forty-one years old, could not have seen how he would, from his mid-sixties until his death at age eighty, be fully concerned with modeling biological theories, and with the creation of a field of biomathematics. Any current search of the biology literature with the keyword "Volterra" will produce literally hundreds of references to Volterra models, the most significant of which are the so-called predator-prey models of interspecies rivalry and population dynamics.

7   Translated as "Mathematical economics in the new *Manual* by Professor Pareto."

8   This was Volterra's "Discorso inaugurale," his inaugural address, at the University of Rome for a chair in mathematical physics. It was initially published in the *Annurio della Università di Roma*, 3–28, and reprinted in the *Giornale degli economisti*, serie 2ª, 23, 436–58 (1901). It was translated into French in 1906, by Ludovic Zoretti, as "Les mathematiques dans les sciences biologiques et sociales" (Volterra 1906). For the reader's conve-

nience, I shall use the English translations, of the relevant portions, done by Giorgio Israel in several of his own papers, though I will on occasion make use of a translation (unpublished) prepared for me by Caroline Benforado. References will be specifically to Israel's translations, as they appeared, with note made of the original source in Volterra either in Italian or French.

9   This specific material is located at the University of California at Berkeley's Bancroft Library in the Griffith Conrad Evans papers, 74/178C, Box 6, Folder "Economics and Mathematics."

10 Let the record show, however, that Evans was in a position to affect the work economists were to do in the postwar period. In a piece on "American Mathematicians in World War I," Price (1988) links Rothrock's (1919) datum that "G. C. Evans of the Rice Institute was a captain of ordinance on special mission in France" to the comment that "since G. C. Evans was president of the [American Mathematical] Society in 1939 and 1940, he participated in the appointment of the War Preparedness Committee of AMS and MAA" (267). We also have the knowledge that "the Aberdeen researchers included such key figures as [Oswald] Veblen, Griffith C. Evans, Marston Morris, Warren Weaver, Norbert Weiner, Hans F. Blickfelt, and G. A. Bliss, the first four of whom played significant roles in mobilizing the country's mathematical expertise during World War II" (Parshall 1994, 444). Thus, Evans was peripherally connected to the emergent cyborg sciences, economics among them, which developed from the collaborative work involving economists and mathematicians during the war (Mirowski 2001).

11 There is thus a delicious irony in the fact that Gerard Debreu had his office at Berkeley in (Griffith Conrad) Evans Hall.

12 "Evans did only a little work in nonequilibrium dynamics. . . . His principal influence upon the progress of economics came through the methodologies employed in his [1930] book, and through his students, among whom were Francis W. Dresch, Kenneth May, C. F. Roos and Ronald W. Shephard, and at one step removed, Lawrence Klein and Herbert A. Simon, who were colleagues or pupils of these students" (Simon 1987, 199).

3   Whose Hilbert?

1   I am grateful to David Reed for an exchange on this point.

2   I think McCloskey rather overstates matters here: "In this paper we develop the mathematical formalism needed to study even-parity perturbations of spherical stellar collapse models, with the goal of calculating the gravitational radiation emitted during a stellar collapse to a black hole or a type II supernova explosion . . . finally, we discuss the numerical solution of this system in a computer code, and various code tests against known results" (Seidel 1990). Physicists use the phrase "mathematical formalism" quite often, and without fear; but they also complain about excessive formalism in papers just as economists do—too much math, not enough physics.

3   Kadish and Tribe (1993, 138) tell us that "criticism of the ideological bias of classical

economic theory often took the form of an attack on abstract theory based purely on deductive reasoning. The 'higher' the theory, the less applicable it became to real conditions. The falsity of Ricardian economics was due not only to its method, but also to its choice of moral premises, thereby implying that the validity of economic theory was necessarily linked to morality. The applicability of any prescriptive policy depended on its practicality and its moral soundness. The Ricardians, on the other hand, had 'set up and worshiped their ill-determined hypothesis of Competition as a natural goal and ideal of social progress.' " Their internal quotation dated from 1885(!), from Oxford, one of the bastions of anti-mathematical economics in England.

4   See, for instance, the very clear discussion of the Hilbert Problems in Grattan-Guinness 2000.

5   This talk was published in *Mathematishe Annalen*, vol. 78, 405–15, 1918. The earlier translation by Fang (1970, 187–98) has been superceded by Ewald (1996, 1105–15); it is this latter to which I shall refer.

6   In fact, another paper of 1922, "Neubegründer der Mathematik. Erste Mitteilung" is, jointly with the 1918 paper, considered to define the program by those who argue that there was indeed such a program. A translation of it as "The New Grounding of Mathematics. First Report" appears in Ewald 1996, 1115–34.

7   I have deliberately avoided calling this program something like "Hilbert's Formalist Program," or "The Formalist Program," or the like. As we shall see later, such labels are quite misleading.

8   In my attempt to be more consistent than those I am here criticizing, I eschew the label "Formalist" for Hilbert's "Programs." One alternative, of course, would be the "Strong Axiomatic Program" and the "Weak Axiomatic Program," though this is not exactly accurate. I present a second alternative below.

9   Some of the material in this section is based on the various papers in the 1988 book *Gödel's Theorem in Focus,* edited by Stuart G. Shanker.

10 I have written about this set of overlapping groups before, with the focus on Abraham Wald (Weintraub 1985, 62–73).

11 In *Monatshefte für Mathematik und Physik*, 38, 173–1998.

12 I note too the recently published biography of John von Neumann by the former editor of *The Economist,* Norman Macrae (1992). This work is extremely thin on von Neumann's connection to economics, and is woefully documented. Further, its concern to glorify von Neumann makes his mathematical activity incomprehensible in its quest to build anecdote on anecdote. Von Neumann's full biography thus remains to be written, although economists do now have a remarkably able portrait of both the man and his concerns in a wonderful chapter in Mirowski's new book *Machine Dreams* (2001), as well the Vienna material (especially) in Leonard (forthcoming).

13 This subject has been the topic of a remarkably adept set of papers on game theory's history by Robert Leonard (1992; 1994; 1995). His book-length treatment of von Neu-

mann and game theory and related matters is an exceptional history of these matters. Philip Mirowski's book (2001) likewise develops the theory of games from many of the same sources that Leonard exploits, and provides a second, remarkably competent history of these matters. My own understanding as well as exposition of these issues has itself developed from discussions with Leonard and Mirowski over several years.

14 "My personal opinion, which is shared by many others, is, that Gödel has shown that Hilbert's program is essentially hopeless" (von Neumann 1947, 231).

15 The best recent discussion of these matters may be found in Smith 1997.

4   Bourbaki and Debreu

This chapter is a revised version of a paper done originally with Philip Mirowski, and which appeared in *Science in Context* (Weintraub and Mirowski 1994).

1 This set of issues has been taken up directly in a remarkable paper by David Aubin (1997).

2 I follow the convention, long-established among mathematicians, and use the masculine "he" instead of a collective pronoun to refer to Nicholas Bourbaki.

3 The original tape recordings of this interview have been deposited with the Economists' Papers Project in the Special Collection Library of Duke University.

4 Transcript prepared, in the first instance, by Ms. Shannon N. Valentine (Durham, North Carolina).

5   Negotiating at the Boundary (with Ted Gayer)

1 The discussion is based on material in the Don Patinkin Papers, located in the Special Collections Library at Duke University.

2 We are extremely grateful to Professor Paul Ehrlich for providing much of the information on Phipps. This history is published on the WorldWide Web at http://tortoise .math.ufl.edu/~theral/mathhist.html. All cites to this are unpaginated.

3 According to his son, Lytle went on to teach those subjects at Florida Tech. His specialty appears to have been Monte Carlo analysis (Lytle 1998).

4 There are many other echoes of Phipps within Miller's dissertation. For example, Miller describes the difference between dependent and independent functions (33–35), which Phipps discusses in his correspondence with Patinkin. Also contained in Miller's dissertation (51) is Phipps's critique of Tintner (1948), which we will discuss in the next section. Most striking is the similar tone of displeasure the two mathematicians share concerning the use of mathematics in economics.

5 We find further evidence of Phipps's cantankerous personality from a letter we received, via Paul Ehrlich, from Professor Tilley, one of Phipps's former students. Tilley relates Phipps's displeasure with Tilley's explanation of the Fundamental Theorem of Calculus

during his Ph.D. oral exams. Tilley writes, "The next day I went to his office and asked him what was his statement of the theorem. He had a stack of about 8 new Calculus texts on his desk. He told me that he would show me the theorem from one of them. He took the top one off and looked up the theorem. He read it to himself and tossed the book into the waste can. He did that with each of the remaining books! He was not satisfied with any of the statements of the theorem. I never did find out exactly what he wanted as he was so unhappy with those texts that he walked out of his office after throwing them all in the trash" (personal communication 1996).

6   The articles he criticized were Patinkin (1948), Tintner (1948), and Friedman (1952a).

7   These early views on models prefigure some interesting recent work by philosophers and historians of economics (see Morrison and Morgan 1998). Morgan's discussion there opens up a new set of issues concerning modeling itself as a compelling objective for economic analysis, as she argues that economics is not well characterized by the division between theory and applications, but rather by modeling in all its complexity.

8   According to Kuhn's view, episodes of paradigm shifts lead to partial breakdowns of communication between the proponents of different theories. In *The Structure of Scientific Revolutions*, Kuhn writes that "the proponents of competing paradigms practice their trades in different worlds. . . . [They] see different things when they look from the same point in the same direction. . . . Both are looking at the world, and what they look at has not changed. But in some areas they see different things, and they see them in different relations on to the other" (150). As mathematics and economics are different disciplines, not simply alternate paradigms within a discipline, it is not Kuhn but rather Stanley Fish's (1980) modification of Kuhnian incommensurability that we adapt to our argument.

9   All references to the Patinkin-Phipps correspondence shall be understood to be from The Don Patinkin Papers, in the Special Collections Library of Duke University.

10 Phipps once initiated a correspondence with Nicholas Georgescu-Roegen concerning what Phipps perceived to be the use of faulty mathematics in Georgescu-Roegen's work. Georgescu-Roegen ended the correspondence after two letters.

11 Of course the relationships between Von Neumann and Morgenstern, as well as Savage and Friedman, offer other cases (with possibly different implications) of communication between mathematicians and economists.

12 Or is said to have claimed. See the discussion of the Thom "claim" in Woodcock and Davis 1978, 70.

6   Equilibrium Proofmaking (with Ted Gayer)

1   The first such history was provided in Weintraub (1983). Related material was developed by Ingrao and Israel (1990 [1987]).

2   Indeed, the philosopher Imre Lakatos (1970) made such "incontrovertibles" the cor-

nerstone of his methodology of scientific research programs, associating them with the "hard core" of the scientific research program; earlier Thomas Kuhn (1962) had of course used a related idea in developing the paradigms of normal science.

3   The image of the "black box" is taken from Bruno Latour, particularly his use of it in Latour 1987. He describes how "black boxes are used by cyberneticians whenever a piece of machinery or a set of commands is too complex. In its place they draw a little box about which they need to know nothing but its input and output" (2–3).

4   Among students taking an early course in econometrics matters were a bit clearer: "What Hercules will attempt to solve the system of equations which we have established above for the determination of general equilibrium! It will be observed that the system is not linear in character . . . [and] it is difficult to believe that the system of demand functions . . . is essentially linear. . . . A partial existence theorem has . . . been given for the mathematical problem by A. Wald, who considered a system of the Walrasian form. He enumerated a set of conditions both on the demand functions and the technical coefficients, which would assure the existence of a mathematical solution of the equations. The complex character of the problem makes it impossible to summarize the analysis here" (Davis 1941, 186–87).

5   "Dear Professor Weintraub: Mr. Horsch has asked me to send you the list of fall adoptions on *Price Theory*. They are as follows: Alabama Polytechnical Institute, 7; University of California at Los Angeles, 55; University of Chicago, 131; Northwestern University, 9; Roosevelt College, 8; University of Delaware, 15; University of Michigan, 6; University of Detroit, 16; Lincoln University, 6; New York University, 25; University of Pittsburgh, 34; Pennsylvania State College, 15; University of Pennsylvania, 13; University of Texas, 19" (letter from Pitman Publishing Corporation, 17 October 1950).

6   As noted in Weintraub (1983), the general economics journal *Zeitschrift für Nationalökonomie* contained a survey piece by Wald in 1936, translated into English and published in *Econometrica* in 1951. Von Neumann's 1936 paper appeared in an English translation in the *Review of Economic Studies* in the 1945–1946 volume. Moreover, Wald's and von Neumann's work was certainly discussed in Schumpeter's graduate economic theory class at Harvard in the late 1930s (Weintraub 1997).

7   This "Chicago" view was in print earlier with Milton Friedman's (1946) hostile review of Oscar Lange's *Price Flexibility and Unemployment*, with Lange of course representing the "other Chicago" of the Cowles Commission. Chicago was the Marshallian antagonist to Cowles's Walrasian predilection.

8   More than a decade earlier in 1941, Oskar Morgenstern had published a hostile review of Hicks's (1939) *Value and Capital* in Chicago's own *Journal of Political Economy*, a piece that effectively sneered at equation-counting to establish equilibrium.

9   The third edition of Stigler's *The Theory of Price* appeared in 1966, well after the publication of Arrow and Debreu's equilibrium proof. Nonetheless, this edition made no mention of Arrow and Debreu, nor did it mention general equilibrium (the chapter on

general equilibrium from the previous edition was dropped). Instead Stigler, continuing to reflect the long-standing Chicago pro-Marshall, anti-Walras position, used partial equilibrium exclusively. A fourth edition of *The Theory of Price* appeared in 1987, again with no mention of general equilibrium. This edition does include a photograph of Arrow, however, and part of the caption mentions Arrow's "fundamental work on the existence of competitive equilibria" (251).

10   In their 1980 third edition textbook, Henderson and Quandt split the "Multimarket Equilibrium" chapter into two chapters: "Multimarket Equilibrium," and "Topics in Multimarket Equilibrium." Nonetheless, the exposition on the existence of an equilibrium is the same as in the previous edition.

11   The following were residents of the Cowles Commission at some point between 1950 and 1953 (see Hildreth 1986): Stephen G. Allen, Kenneth J. Arrow, Pierre F. J. Baichere, Earl F. Beach, Gary S. Becker, Martin J. Beckmann, Francis Bobkoski, Karl Borch, George H. Borts, Karl Brunner, Rosson L. Cardwell, Herman Chernoff, John Chipman, Carl F. Christ, Gerard Debreu, William L. Dunaway, Atle Harald Elsas, Karl Fox, Jose Gil-Pelaez, Thomas A. Goldman, William Hamburger, I. N. Herstein, Clifford Hildreth, William C. Hood, Henry S. Houthakker, Leonid Hurwicz, Herman F. Karreman, Tjalling C. Koopmans, Jules Levengle, Siro Lombardini, C. B. McGuire, Pierre Maillet, Edmond Malinvaud, Sven Malmquist, Harry Markowitz, Jacob Marschak, Rene Montjoie, Marc Nerlove, William Parrish, Sigbert J. Prais, Roy Radner, Stanley Reiter, Bertram E. Rifas, Herman Rubin, Sam H. Schurr, William B. Simpson, Morton L. Slater, Gerhard Stoltz, Erling Sverdrup, James G. Templeton, Ciro Tognetti, Leo Tornqvist, Jaroslav Tuzar, Daniel Waterman, Isamu Yamada, and Jagna Zahl.

12   The correspondence on which this material is based is preserved in the Nicholas Georgescu-Roegen Papers, located in the Special Collections Library at Duke University. Nicholas Georgescu-Roegen, a Romanian-born economist, was an associate editor at *Econometrica* during the period we are considering. Georgescu-Roegen studied mathematics at the University of Bucharest and earned his Ph.D. in statistics at the Sorbonne in 1932, and studied for a period under the statistician Karl Pearson at the University College in London. He visited the United States in 1934, where he became interested in economics due to the influence of Joseph Schumpeter. During this time he published his influential paper "The Pure Theory of Consumer Behavior" in the *Quarterly Journal of Economics*. He returned to Romania in 1936, but came to America for good in 1948, and spent most of his career at Vanderbilt University. In 1971 he published his book *The Entropy Law and the Economic Process*, in which he claimed that the second law of thermodynamics implies that economic processes lead the world toward disorder, and thus the steady-state equilibrium commonly ascribed to by neoclassical economics is impossible.

13   This of course is part of the larger story of the creation of a community of mathematically adept social scientists in the wartime and immediate postwar period. This story is well told in Mirowski (2001) and in Leonard (forthcoming).

14 William J. Baumol, born in New York City, received his BSS from College of the City of New York in 1942. He received his doctorate from the University of London in 1949, where he wrote a dissertation on *Welfare Economics and the Theory of the State*. He taught at the London School of Economics from 1947 to 1949, then left to join the faculty at Princeton. His book *Economic Dynamics* established him as a mathematically able economic theorist. He became a full Professor at Princeton in 1954, and since 1971 he has held a joint appointment there and at New York University where he pioneered a new area of study, the economics of the arts.

15 The paper states, immediately following the statement of the lemma, that it "generalizes Nash's theorem on the existence of equilibrium points for games."

16 A recent study by Hamermesh (1994) examines what characteristics editors look for in referees. He suggests that the current practice of top journals is to use heavily cited people to serve as referees, especially when the author is well known. In this context, the choice of Phipps as a referee is odd. However, Hamermesh also finds that journal editors frequently choose as referees people who have recently published an article in the journal. Phipps had two *Notes* that appeared in *Econometrica* in 1950.

17 We note that some of the exact language—wording and phrasing—in the Miller thesis appears as Phipps's words in letters we have from Phipps to Don Patinkin (chapter 5).

18 We have been unable to locate any corroborating biographical material on Miller. He does not appear to have left traces in any literature, or to have been noted in any material uncovered by Paul Ehrlich in his history of mathematics at the University of Florida, or anyone's memories at Clemson. Our only knowledge of Miller thus comes from his unpublished doctoral dissertation, where the "Vitae" on the last page tells us that Miller was born in 1911 in Birmingham, Alabama, and graduated from Birmingham-Southern College in 1931. He received his mathematics M.A. from Florida in 1933, and then taught high school and coached sports in Alabama. He got a job as an engineer in Pittsburgh in 1936, and in 1938 he became an Instructor in Mathematics at Clemson College. Following military service, he returned to Clemson, and eventually got a leave of absence to finish his Ph.D. in the period 1949–1951. He was thus forty when he received his degree, and presumably returned to Clemson. To be fair, Phipps was not always so off base. He did publish a couple of correction notes on papers by Milton Friedman (which appeared in the *Journal of Political Economy*) and Gerhard Tintner (which appeared in *Econometrica*), and he published a lengthy paper criticizing Patinkin's monetary theory in *Metroeconomica* (chapter 5).

19 Nonetheless, it is worthwhile to point out that although Georgescu-Roegen was in the habit of writing comments throughout the margins of submitted manuscripts, his copy of Arrow and Debreu's manuscript contains written comments only in the margins of the preliminary sections, with no comments in the margins of the sections containing the proofs.

20 The American Summer Meeting of the Econometric Society was held in Kingston, Ontario, Canada from 31 August–4 September 1953. Among others, attending mem-

bers of the program committee were Debreu and McKenzie. Baumol, Koopmans, Strotz, and Georgescu-Roegen were also present, and McKenzie gave a paper, discussed by Koopmans, on "Competitive Equilibrium with External Economies" in a session chaired by Strotz. Small world indeed.

21   In my earlier history of the theory (Weintraub 1983), I insisted on the term "Arrow-Debreu-McKenzie Model." The usage has not generally taken hold. The connection of Lionel McKenzie's work to that of Arrow and Debreu is a tricky subject to broach as two of the three men have won the Nobel Prize with citations noting the existence proof. McKenzie's work was independently done, and his use of the Kakutani fixed point theorem to prove the existence of the general equilibrium is still the favored expository route, and is the one used later by Debreu in his *Theory of Value*, and by Arrow in his textbook with Frank Hahn, *General Competitive Analysis*. A simple-minded chant of "Mertonian simultaneous discovery" seems not to suffice. After all, in the process we are describing, *Econometrica* referee/editor Georgescu-Roegen asks that the Arrow-Debreu proof be modified to resemble McKenzie's proof!

22   The eighteenth and final item is the set of fourteen minor corrections, like typographical errors, notation confusions, and suggested wording changes.

23   He does state that it "is difficult to see how this solution partakes of the nature of a game." He also criticizes the paper for not considering the uniqueness of the solution.

24   However, other studies (such as Zuckerman and Merton 1971) find no evidence of institutional bias on the part of referees.

25   This kind of self-reflecting, self-aggrandizing referee comment by Frisch has been noted elsewhere. As Samuelson has recalled: "Ragnar Frisch was pretty much autonomous editor of the early issues of *Econometrica*. He was interested in everything. Also, he believed in the superiority of his interpretations of anything and everything (indeed, he was so great a mind that there was much merit in such a belief). When Wassily Leontief participated in the post-1933 revival of the economic theory of index numbers . . . Frisch held up publication of the Leontief 1936 contribution until he could publish in the same issue of *Econometrica* his own survey article on the subject. Foul play, I say" (as quoted in Shepherd 1995, 23).

7   Sidney and Hal

1   Quotations from that autobiography will be taken from the Kregel collection's version, and will be noted as Weintraub 1988 (1983).

2   That typescript, titled "I remember Hal," is written in verse form.

3   Penned literally. Sidney's stationery, and pens, were rough especially during the war, and his handwriting was never very legible. This got worse during the war when he began a private war with U.S. Army mail censors, testing the limits of their vision and patience as he developed a cursive script best characterized as a wavy line with bumps.

Although he never appeared to have been censored (there are no excisions or white-outs in the letters), slow delivery was certainly an issue. I have thus had to make informed guesses at times about words in the letters, and have also corrected obvious spelling and grammar errors. Sidney's writing letters during lights-out, hiding in a latrine, while exhausted, was not conducive to close self-editing. I trust that my own editing of the letters thus will be understood and forgiven.

4   Weintraub's discussion of aggregate supply was really the first full-scale "rehabilitation" of Keynes's own treatment to appear in the Keynesian literatures in the United States, and set out the path trod later by Post Keynesians to criticize the Hicksian IS-LM framework.

5   In addition to the large Post Keynesian literature on this topic, see Weintraub 1979, chapter 3.

6   Weintraub's previous publisher, Pitman of London, apparently had decided to end its economics list, so Weintraub approached American firms like McGraw-Hill, which were not encouraging.

7   Silesia, in central Europe, was a rich farm, factory, and mine (iron, zinc, coal) region divided into German (Upper and Lower) and Austrian Silesia before World War I. After the German defeat in WWII, Austrian Silesia was returned to Czechoslovakia, and nearly all of German Silesia was included in Poland.

8   This was the home of the Brooklyn Dodgers. For the record, there appears to be no connection between Sidney and "Phumbling Phil Weintraub" who played for the New York Giants from 1933 to 1945.

9   This is not the same as making a claim about the sociology of mathematical knowledge. David Bloor, for instance, has some exemplary analysis of the social fabric of mathematical knowledge (Bloor 1991; Barnes, Bloor, and Henry 1996).

10   The only way currently to suggest some answers to this kind of question would be with a systematic study of some comparative autobiographical materials: perhaps one place to start would be to compare similarities and differences found in the two-volume *Mathematical People* with the two volumes of *Recollections of Eminent Economists,* supplemented by other autobiographical writings from mathematicians and economists.

11   One starting place might be with the Klamer and Colander volume on *The Making of an Economist.* One suspects there are some studies by the mathematician's professional organizations on similar topics.

12   For instance, the utter sincerity with which mathematicians believe that a person's intellectual products can be absolutely ranked on invariant scales of quality quite unnerve scholars in other disciplines. I once served on a university tenure committee in which the chair of mathematics was asked where his candidate would place among all current tenured mathematics faculty, a question asked to get an idea whether the average departmental quality would be "improved" by the appointment. The chair replied that the person would rank behind X, Y, and Z, but precisely ahead of A, who was ahead

of B, who ranked ahead of C, and so on for the next fourteen individuals. The historian on the committee had to be convinced that the ranking was not meant as a joke.

13   "My supreme mentor, by sheer accident, was Thomas Francis Patrick McManus. Out of Iowa, he was teaching in the New York suburbs and audited some classes I attended. Soon he gravitated to me as a promising student; he knew well, despite the eminent reputation of many of my teachers, that I was under exposed to the exciting trailblazers of the 1930s. He directed me to Robbins's *Essay*, Keynes's *Treatise*—whose two volumes I outlined, page after uncut page—Hayek, Dennis Robertson, Knight's *Risk*, and *The Ethics of Competition*. (I could later correct Knight—in conversation—when he wrote in a review article that he had never read Wicksteed!) There was also Wicksell, Von Mises's *Socialism* and *Money and Credit*, Chamberlin, Joan Robinson, and Schumpeter's *Theory of Economic Development*—still his most original book, in my view" (Weintraub 1989, 39).

14   Even in London he was quite anti-Roosevelt. Commenting on the much-rumored plan, after Munich, to move the royal family to Canada, he wrote, "If we remove the President to Siberia—a fine place for him, to freeze his smile—he'd no longer be President of the United States. Think of all the income tax people would save" (20 November 1938).

15   Joan Robinson referred to it as "the little green horror," playing on the cover color of that journal.

16   Sraffa was a Fellow of King's College, not actually a professor. Neither of course was Kaldor a professor: John King has reminded me that Kaldor had failed (!) mathematics at LSE as a student.

17   "By the way I don't know what if any news you've had from Machlup. It may well be he's written you of problems and discouragement. If so you can forget to a large extent what he said. I'm convinced of what could be done with the work. Don't forget it or overlook it for I shan't" (20 December 1944). Sidney was later convinced (Weintraub 1988, 46) that Machlup had "borrowed" parts of the manuscript for his own *The Economics of Sellers' Competition; model analysis of sellers' conduct* (1952).

18   From a letter, 13 February 1945, Sidney noted that the unit that he was attached to in London was formally in G-2, the so-called intelligence section, and its full title was the Combined (British and American) Intelligence Objectives Subcommittee. His own job there was to open the envelopes that contained the daily combined report in quadruplicate, place one copy in the British group's mailbox, one copy in the American group's mailbox, one copy in the Subcommittee file, and forward one copy on to the supervising office. This work took all morning each day, leaving him plenty of time for softball.

19   It was a problem for Sidney's appointment that Simon Kuznets was already at Penn (though he was to leave for Johns Hopkins in 1954), as was Irving B. Kravis, because the "gentleman's agreement" in effect at that time was that the department of economics could have no more than two Jewish professors. Sidney made three.

20   Hal had married a former student of his from Jackson College (then Tufts's coordinate

college for women). They married in full awareness of Hal's illness, and with an understanding of the likely limits of time for their life together.

21 Hal's letter continues with an unusual bit of information, given some idea of the person behind these letters: "I had my name and picture on the front page of several Boston papers recently. A friend and myself rescued a 5 year old who had fallen into a ditch in a field across from where we lived. The child was reported lost about 7:00 P.M. We found him at 9:30 P.M. in a cold driving rain, unable to climb up to the top of the ditch. He was completely shocked and muddy. We received news coverage befitting real heroes, though the rescue was very undramatic. Some of the papers provided their own dynamic action."

8   From Bleeding Hearts to Desiccated Robots

1 In 1998 I found, in some of my father's papers, my acceptance letter to Penn, which he had intercepted and withheld from me so that I would not even think of going there as an undergraduate.

2 For example, classmate Duncan Foley was a mathematics major who minored in classics and economics, while my thrice roommate, Peter Weinberger, minored in physics and philosophy before going off, after a Ph.D., to co-create UNIX at Bell Labs.

3 Although physicists themselves were, in their own view, except for us mathematicians, *über alles:* "Every few years this same issue keeps coming up—social "scientists" wanting to be called scientists. Sorry folks, but psychology, sociology, and their offshoots (e.g., ethnic or gender studies) are not and never will be considered a science by those of us who are real scientists. Real science attempts to derive laws that can explain observed phenomena. This derivation process involves two steps: formulation of a theory and the presentation of evidence that validates that theory. This validation must either be a formal mathematical proof, or it must be supported by experimental results. Since there is not a single example of a formal mathematical proof in any social science, I am forced to focus only on the experimental evidence validation method. Experimental data must be collect from a credible experiment that—and this is key—must be repeatable by other competent researchers. Furthermore, this data must be analyzed either mathematically or by conducting additional experiments. Much of social science research amounts to little more that collecting huge amounts of demographic data which is then subjected to a statistical analysis to look for trends. Unfortunately, the social science researcher has little or no control over the source of the data. Hence, there is little opportunity to guarantee data integrity thereby tainting any conclusions that are drawn. For example, census data is used in many sociological investigations even though many groups complain they are underrepresented. If their claims are true, what does this say about any conclusions based on a statistical analysis of that census data? What can any social scientist do to correct this situation? My purpose is not to cast

aspersions at the sociologists. But sociology (and other social science fields) does not have any laws. They only have personal theories, which are supported by data subject to interpretation rather than a rigorous analysis.

    I once had a professor tell me that any academic field which has the word "science" as part of its name was not a true science. Examples abound (political science, family and consumer sciences, etc.). Social science certainly fits into this category. Let's reserve the word scientist for those who are really conducting scientific work.—Carol Amadu, Independent Scholar (in physics), University of Maryland (posted to "Colloquy," discussion group of *Chronicle of Higher Education* 8/7, 6:00 P.M., U.S. Eastern time).

4   I had but one introductory economics course at Swarthmore, which I recall only dimly, and in which I received grades of C+ and B- in the two semesters.

9 Body, Image, and Person

1   For Galison, these involve "problems of pidginization and creolization. Both refer to language at the boundary between two groups. A pidgin usually designates a contact language constructed with the elements of at least two active languages. . . . A creole, by contrast, is by definition a pidgin extended and complexified to the point where it can reasonably serve as a stable native language" (Galison 1999 [1997], 153–54).

2   It should not be a surprise at this point in the book for the reader to appreciate that Sidney Weintraub, increasingly marginalized in the Economics Department at the University of Pennsylvania, found access to graduate students primarily by teaching the history of economics. His approach was through the canonical texts (Smith, Malthus, Ricardo, etc.) to show that economics was characterized by a set of progressive moves until Keynes's *General Theory,* and degenerating moves ever since that time.

3   Many historians of science today would argue that that experiment was irrelevant to the theoretical developments in physics, its importance being rather a narrative importance in its fitness for the story of falsificationism as good science.

4   For a remarkable reinterpretation of Lakatos's ideas, as less Popperian than Hegelian, based on new evidence of his political background in Hungary, see Kadvany's *The Guises of Reason* (2001).

5   For example, it proved necessary to recast the labor theory of value.

6   For the record, Kuhn thought not, arguing himself that neither economics, nor any other social science, were real sciences in his sense of the term. Ultimately, that demarcation issue of separating science from non-science is irrelevant for us, as we more simply ask whether the Kuhnian framework structures an interesting narrative, and if so what are the characteristics of that narrative.

7   The loveliest examples of this phenomenon can be found in autobiographical accounts by economists (see Szenberg 1992).

8   For an excellent example of this tension, see Paul Samuelson's "Out of the Closet: A

Program for the Whig History of Economic Science" (1987), which explicitly repudiates the idea of historical reconstruction.

9   Newton of course was proffering a modest self for public consumption, quite at odds with his private persona (Westfall 1980). The claim by Samuelson that he too "stood on the shoulders of giants" (Samuelson 1983, xxiv–xxv) can then be read as offering a similar modest person for public consumption, but underneath of course lies the reader's knowledge that the claim equates Samuelson with Newton, rather a stretch, it would appear.

10 For Latour (1987), networks are linked sets of actants, where actants are both human and nonhuman. The biology bench-scientist is an actant, as is the microorganism she claims she observes with her instrument actant, etc. They are all actors in the network.

11 This argument, which to me appears unassailable, is well made in various pragmatist literatures. See, for example, "Introduction: Going Down the Anti-Formalist Road" (Fish 1989), or Novick 1988.

12 Robert Skidelsky's magnificent biography of Keynes (1986) must be noted, but Skidelsky is not an economist, while Dierdre McCloskey's (1999) recent autobiographical memoir is not much addressed to economists.

Bibliography

Houston Chronicle. 1915a. "Scientific Aspects of Philosophy." 6 May.

——. 1915b. "Scientific Aspects of Philosophy." 13 May.

——. 1916. "Scientific Aspects of Philosophy." 28 January.

——. 1933. "Dr. G. C. Evans resigns from Rice Institute." 27 October, 1–2.

——. 1973. "Services Held For Dr. Griffith Evans." 10 December, 3, 25.

Houston Post. 1916. "Rice Professors Honored By Northern Institutions." 9 January.

——. 1918. "Rice Professor Attaché of Embassy at Rome." 30 December.

——. 1919. "Dr. Evans at Washington after Service in France." 5 June.

——. 1925. "Rice 'Prof.' Is Honored By Math Society." 1 February.

Abrams, M. H. 1972. "What's the Use of Theorizing About the Arts?" In *In Search of Literary Theory*, edited by M. Bloomfield. Ithaca, NY: Cornell University Press.

Albers, D. J., and C. Reid. 1990. "Conversation With Steve Smale (October 1985)." In *More Mathematical People: Contemporary Conversations* by D. J. Albers, G. L. Alexanderson, and C. Reid. New York: Harcourt Brace Jovanovich: 304–23.

Allais, M. 1943. *A la Recherche d'une discipline economique. Premiere partie: l'economie pure.* Paris: Ateliers Industria.

Allen, R. D. G. 1931. "Review of Mathematical Introduction to Economics by Evans." *Economica, Old Series* 11(31): 108–9.

——. 1938. *Mathematical Analysis for Economists.* London: Macmillan.

Andler, M. 1988. "Entretien avec trois membres de Nicholas Bourbaki." *Gazette des Mathématiciens* 35: 43–49.

Arrow, K. J. 1954. "A Review of 'Money in the Utility Function.'" *Mathematical Reviews* 18: 18366h.

Artin, E. 1953. "Review of Bourbaki [Elements, Book II, Chapters 1–7 1942–1952)]." *Bulletin of the American Mathematical Society* 59: 474–479.

Ashmore, M., et al. 1994. "The Bottom Line: The Rhetoric of Reality Demonstrations." *Configurations* 2(1): 1–14.

Atiyah, M., et al. 1994. "Responses to 'Theoretical mathematics: Toward a Cultural Synthesis of Mathematics and Theoretical Physics.' " *Bulletin of the American Mathematical Society (N.S.)* 30(2): 178–207.

Aubin, D. 1997. "The Withering Immortality of Nicholas Bourbaki: A Cultural Connector at the Confluence of Mathematics, Structuralism, and the Oulipo in France." *Science in Context* 10(2): 297–342.

Bachelard, G. 1984 [1934]. *The New Scientific Spirit*. Boston: Beacon Press.

Backhouse, R. 1992a. "The Constructivist Critique of Economic Methodology." *Methodus* 4(2).

—. 1992b. "How Should We Approach the History of Economic Thought: Fact, Fiction, or Moral Tale?" *Journal of the History of Economic Thought* 14(1): 18–35.

Bagemihl, R. 1958. "Review of Bourbaki [Elements, Book I, Chapter 3]." *Bulletin of the American Mathematical Society* 64: 390.

Bagioli, M., ed. 1999. *The Science Studies Reader*. London: Routledge.

Ball, W. W. R. 1889. *A History of the Study of Mathematics at Cambridge*. Cambridge: Cambridge University Press.

Barnes, B., et al. 1996. *Scientific Knowledge: A Sociological Analysis*. Chicago: University of Chicago Press.

Barrington, J. 1981. "15 New Ways to Catch a Lion." In *The Best of Manifold: 1968–1980*, edited by I. Stewart and J. Jaworski. Nantwich, Cheshire: Shiva Publishing.

Barwise, J., and J. Moravcsik. 1982. "A Review of Formal Philosophy. The Selected Papers of Richard Montague, by Richmond H. Thomason." *Journal of Symbolic Logic* 47(1): 210–215.

Baumol, W. J., and S. M. Goldfeld, eds. 1968. *Precursors in Mathematical Economics*. LSE Reprints of Scarce Works on Political Economy, No. 19. London: London School of Economics.

Birkhoff, G. D. 1933. *Aesthetic Measure*. Cambridge, MA: Harvard University Press.

Black, R. D. C., et al., eds. 1973. *The Marginal Revolution in Economics*. Durham, NC: Duke University Press.

Blaug, M. 1980. *The Methodology of Economics*. New York: Cambridge University Press.

—. 1999. "The Formalist Revolution or What Happened to Orthodox Economics after World War II?" In *From Classical Economics to the Theory of the Firm*, edited by R. E. Backhouse and J. Creedy. Cheltenham, UK: Edward Elgar.

Bloor, D. 1991. *Knowledge and Social Imagery*. Chicago: University of Chicago Press.

Boi, L., et al., eds. 1992. *1830–1930: A Century of Geometry: Epistemology, History and Mathematics*. Lecture Notes in Physics. Berlin: Springer-Verlag.

Boles, R. C. 1998. Telephone conversation with E. Roy Weintraub, 2 January 1998.

Bourbaki, N. 1949. "The Foundations of Mathematics for the Working Mathematician." *Journal of Symbolic Logic* 14(1): 1–8.

—. 1950. "The Architecture of Mathematics." *American Mathematical Monthly* 57: 221–232.

Bowker Co. 1955. *American Men of Science*. New York: R. R. Bowker Co.

Bowley, A. L. 1932. "Review of Mathematical Introduction to Economics by Griffith C. Evans." *The Economic Journal* 42(165): 93–94.

Brandt, L. 1959. "Rice Educator to Entertain." *Houston Chronicle*. 9, 2.

Breit, W., and R. W. Spencer 1995. *Lives of the Laureates*. Cambridge: MIT Press.

Brody, A. 1989. "Economics and Thermodynamics." In *John von Neumann and Modern Economics*, edited by M. Dore, S. Chakravarty, and R. Goodwin. Oxford: Clarendon Press.

Bronfenbrenner, M. 1987. "A Conversation with Martin Bronfenbrenner." *Eastern Economic Journal* 13(1) (January–March): 1–6.

Browder, F. E. 1988. "Mathematics and the Sciences." In *History and Philosophy of Modern Mathematics*, vol. 9, edited by W. Aspray and P. Kitcher. Minneapolis: University of Minnesoata Press, 278–92.

——. 1989. "The Stone Age of Mathematics on the Midway." In *A Century of Mathematics in America. Part II*, edited by P. Duren. Providence: American Mathematical Society.

Cambridge University. 1885. *Ordinances of the University of Cambridge*. Cambridge: Cambridge University Press.

Campbell, D. M., and J. C. Higgens, eds. 1984. *Mathematics: People, Problems, Results*. Belmont, CA: Wadsworth.

Carpenter, E. 1916. *My Days and Dreams (Being Autobiographical Notes)*. New York: Charles Scribner's Sons.

Cartan, H. 1943. "Sur le fondement logique des mathématiques." *La Revue Scientifique: Revue rose illustrée* 81(1): 3–11.

——. 1980. "Nicholas Bourbaki and Contemporary Mathematics." *The Mathematical Intelligencer* 2(4): 175–180.

Cassidy, J. 1996. "The Decline of Economics." *New Yorker*: 50–60.

Caws, P. 1988. *Structuralism: The Art of the Intelligible*. London: Humanities Press.

Chadarevian, S. D. 1997. "Using Interviews to Write the History of Science." In *The Historiography of Contemporary Science and Technology*, edited by T. Söderqvist. Amsterdam: Harwood.

Chikara, S., et al., eds. 1994. *The Intersection of History and Mathematics*. Science Networks—Historical Studies. Basel: Birkhäuser Verlag.

Christ, C. 1952. *Economic Theory and Measurement*. Chicago: Cowles Commission.

Clarke, P. 1998. *The Keynesian Revolution and Its Economic Consequences*. Cheltenham, UK: Edward Elgar.

Coats, A. W. 1993. *The Sociology and Professionalization of Economics*. London: Routledge.

——, ed. 1996. *The Post-1945 Internationalization of Economics*. Durham, NC: Duke University Press.

Collins, H. 1985. *Changing Order*. Beverly Hills, CA: Sage.

Cook, S. 2000. Letter to E. Roy Weintraub.

Corry, L. 1989. "Linearity and Reflexivity in the Growth of Mathematical Knowledge." *Science in Context* 3(2): 409–40.

—. 1992. "Nicholas Bourbaki and the Concept of Mathematical Structure." *Synthese* 92(3): 315–48.

—. 1996. *Modern Algebra and the Rise of Mathematical Structures*. Boston, Birkhäuser.

—. 1997. "David Hilbert and the Axiomatization of Physics (1894–1905)." *Archive for History of Exact Sciences* 51: 88–197.

Cournot, A. 1963. *Researches into the Mathematical Principles of the Theory of Wealth*. Homewood, IL: Irwin.

Craver, E. 1986. "The Emigration of the Austrian Economists." *History of Political Economy* 18(1): 1–36.

Crawford, T. H. 1992. "An Interview with Bruno Latour." *Configurations* 1(2): 247–269.

—. 1994. "Review of Bruno Latour's 'We Have Never Been Modern.'" *Configurations* 2(3): 578–80.

Darnell, A. C. 1990. "Introduction: The Life and Economic Thought of Harold Hotelling." In *The Collected Economic Articles of Harold Hotelling* by A. C. Darnell. New York: Springer-Verlag: 1–35.

Dauben, J. 1992. "Are There Revolutions in Mathematics?" In *The Space of Mathematics,* edited by J. Echeverria, A. Ibarra, and T. Mormann. Berlin: Walter de Gruyter.

—. 1994. "Mathematics: An Historian's Perspective." In *The Intersection of History and Mathematics,* edited by S. Chikara, S. Mitsuo, and J. W. Dauben. Basel: Birkhäuser Verlag. 15: 1–14.

Davis, H. T. 1941. *The Theory of Econometrics*. Bloomington, IN: Principia Press.

Davis, P. J., and R. Hersh 1981. *The Mathematical Experience*. Boston: Birkhäuser.

—. 1986. "The Ideal Mathematician." In *New Directions in the Philosophy of Mathematics,* edited by T. Tymoczko. Princeton: Princeton University Press. 177–84.

Dawson, J. 1979. "The Gödel Incompleteness Theorem from a Length-of-Proof Perspective." *American Mathematical Monthly* 86(9): 740–47.

—. 1988a. "Kurt Gödel in Sharper Focus." In *Gödel's Theorem in Focus,* edited by S. Shanker. Beckenham, UK: Croom Helm. 1–16.

—. 1988b. "The Reception of Gödel's Incompleteness Theorem." In *Gödel's Theorem in Focus,* edited by S. Shanker. Beckenham, UK: Croom Helm. 74–95.

Dear, P. 1995. *Discipline and Experience: The Mathematical Way in the Scientific Revolution.* Chicago: University of Chicago Press.

Debreu, G. 1959. *The Theory of Value*. New York: John Wiley.

—. 1983. *Mathematical Economics: Twenty Collected Papers of Gerard Debreu*. New York: Cambridge University Press.

—. 1984. "Economic Theory in the Mathematical Mode." *American Economic Review* 74(3): 267–78.

—. 1986). "Theoretical Models: Mathematical Form and Economic Content." *Econometrica* 54(6): 1259–70.

—. 1991. "The Mathematization of Economic Theory." *American Economic Review* 81(1): 1–7.

—. 1992). "Random Walk and Life Philosophy." In *Eminent Economists: Their Life Philosophies,* edited by M. Szenberg. New York: Cambridge University Press. 107–14.

DeMarchi, N. B., ed. 1993. *Non-Natural Social Science: Reflecting on the Enterprise of More Heat Than Light.* Durham, NC: Duke University Press.

DeMarchi, N. B., and M. Blaug, eds. 1991. *Appraising Economic Theories: Studies in the Methodology of Research Programs.* Aldershot, UK: Edward Elgar.

Dennis, K. 1995. "A Logical Critique of Mathematical Formalism in Economics." *Journal of Economic Methodology* 2(2): 181–99.

Dieudonne, J. 1939. "Les methodes axiomatiques modernes et les fondements des mathématiques." *La Revue Scientifique: Revue rose illustrée* 77(3): 224–32.

—. 1970. "The Work of Nicholas Bourbaki." *American Mathematical Monthly* 77: 134–45.

—. 1982a. *A Panorama of Pure Mathematics (as seen by Nicholas Bourbaki).* New York: Academic Press.

—. 1982b. "The Work of Bourbaki in the Last Thirty Years." *Notices of the American Mathematical Society* 29: 618–23.

Douglas, M. 1989 [1966]. *Purity and Danger: An Analysis of Concepts of Pollution and Taboo.* London: Ark.

Dow, S. C. 1985. *Macroeconomic Thought: A Methodological Approach.* Oxford: Basil Blackwell.

Dreze, J. 1964. "Some Postwar Contributions to French Economists to Theory and Public Policy, with Special Emphasis on Problems of Resource Allocation." *American Economic Review* 54 (4, Part 2—Supplement): 1–64.

Eatwell, J., et al., eds. 1987. *The New Palgrave: A Dictionary of Economics.* New York: Stockton Press.

Echeverria, J., et al., eds. 1992. *The Space of Mathematics: Philosophical, Epistemological, and Historical Explorations.* Berlin: Walter de Gruyter.

Edgeworth, F. Y. 1889. "Points at Which Mathematical Reasoning is Applicable to Political Economy." *Nature* 40 (September 19): 496–501.

—. 1985 [1881]. *Mathematical Psychics and Other Essays.* Mountain Center, CA: James and Gordon.

Eilenberg, S. 1942. "Review of Bourbaki [Elements, Book I, Fascicule des résultats]." *Mathematical Reviews* 3: 55.

—. 1945. "Review of Bourbaki [Elements, Book II, Chapter 1]." *Mathematical Reviews* 6: 113.

Elkana, Y. 1981. "A Programmatic Attempt at an Anthropology of Knowledge." In *Sciences and Cultures,* edited by E. Mendelsohn and Y. Elkana. Dordrecht: Reidel.

Epstein, R. J. 1987. *A History of Econometrics.* Amsterdam: North-Holland.

Evans, G. C. 1920. "Fundamental Points of Potential Theory." *Rice Institute Pamphlet* 7(4): 252–329.

—. 1921. "The Physical Universe of Dante." *Rice Institute Pamphlet* 8(2): 91–117.

—. 1922. "A Simple Theory of Competition." *American Mathematical Monthly* 29: 371–80.

——. 1924. "The Dynamics of Monopoly." *American Mathematical Monthly* 31: 77–83.

——. 1925. "Economics and the Calculus of Variations." *Proceedings of the National Academy of Sciences* 11(1): 90–95.

——. 1925. "The Mathematical Theory of Economics." *American Mathematical Monthly* 32: 104–10.

——. 1926. "The Place of Francis Bacon in the History of Scientific Method." *Rice Institute Pamphlet* 13(1): 73–92.

——. 1929. "Cournot on Mathematical Economics." *Bulletin of the American Mathematical Society* 35 (March–April): 269–71.

——. 1930. *Mathematical Introduction to Economics*. New York: McGraw-Hill.

——. 1931. "A Simple Theory of Economic Crises." *American Statistical Association Journal, Supplement* 26 (March): 61–68.

——. 1932. "The Role Of Hypothesis in Economic Theory." *Science* 75 (1943): 321–24.

——. 1934. "Maximum Production Studied in a Simplified Economic System." *Econometrica* 2 (1): 37–50.

——. 1959. Preface to the Dover Edition. *Theory of Functionals and of Integral and Integro-differential Equations* by V. Volterra. New York: Dover.

——. 1967. Autobiographical Fragments. 74/178C, Box 6 Folder "Evans, G. C. Biography," Manuscript Collection, Bancroft Library, University of California at Berkeley.

Ewald, W., ed. 1996. *From Kant to Hilbert: A Source Book in the Foundations of Mathematics: Volume II*. Oxford: Oxford University Press.

Ewing, J. 1992. "Review of The History of Modern Mathematics." *Historia Mathematica* 19(1): 93–98.

Fang, J. 1970. *Hilbert: Towards a Philosophy of Modern Mathematics II*. Hauppage, NY: Paideia Press.

Fay, B., et al., eds. 1998. *History and Theory: Contemporary Readings*. Oxford: Blackwell.

Feiwel, G. 1987a. "Oral History II: An Interview with Gerard Debreu." In *Arrow and the Ascent of Modern Economic Theory*, edited by G. Feiwel. New York: New York University Press. 243–57.

Feiwel, G., ed. 1987b. *Arrow and the Ascent of Modern Economic Theory*. New York: New York University Press.

Fish, S. 1980. *Is There a Text In This Class?* Cambridge, MA: Harvard University Press.

——. 1989. *Doing What Comes Naturally*. Durham, NC: Duke University Press.

——. 1995. *Professional Correctness*. Oxford: Clarendon Press.

——. 1997. "Mission Impossible: Settling The Just Bounds Between Church And State." *Columbia Law Review* 97(8): 2255–333.

Fisher, C. S. 1984 [1972/3]. "Some Social Characteristics of Mathematicians and Their Work." In *Mathematics: People, Problems, Results*. Volume III, edited by D. M. Campbell and J. C. Higgens. Belmont, CAL Wadsworth. 230–47.

Fisher, I. 1892. *Mathematical Investigations into the Theory of Value and Prices*. New Haven: Connecticut Academy of Arts and Sciences.

Fleck, L. 1979 [1935]. *Genesis and Development of a Scientific Fact*. Chicago: University of Chicago Press.

Forsyth, A. R. 1935. "Old Tripos Days in Cambridge." *Mathematical Gazette* 19: 162–79.

Fox, K. A. 1987. "Roos, Charles Frederick." In *The New Palgrave: A Dictionary of Economics*, edited by J. Eatwell, M. Milgate, and P. Newman. New York: Stockton Press. 4: 219–20.

Friedman, M. 1952a. "A Reply." *Journal of Political Economy* 60: 334–36.

——. 1952. "The 'Welfare' Effects of an Income Tax and an Excise Tax." *Journal of Political Economy* 60: 25–33.

Frink, O. 1950. "Review of Bourbaki 1949." *Mathematical Reviews* 11: 73.

Galison, P. 1987. *How Experiments End*. Chicago: University of Chicago Press.

——. 1997. *Image and Logic: A Material Culture of Microphysics*. Chicago: University of Chicago Press.

——. 1999 [1997]. "Trading Zone: Coordinating Action and Belief." In *Science Studies Reader*, edited by M. Bagioli. New York: Routledge.

Gandolfo, G. 1987. "Volterra, Vito." In *The New Palgrave: A Dictionary of Economics*, edited by J. Eatwell, M. Milgate, and P. Newman. New York: Stockton Press. 4: 817–18.

Gandy, R. O. 1959. "Review of Bourbaki." *Journal of Symbolic Logic* 24: 71–73.

Gayer, T. 1997. "Hazardous Waste Risks, Housing Prices, and Economic Methodology." Duke University Economics Department.

Geertz, C. 1988. *Works and Lives*. Stanford: Stanford University Press.

Gell-Mann, M. 1992. "Nature Conformable to Herself." *Bulletin of the Santa Fe Institute* 7(1): 7–10.

Gilbert, G. 1991. "La scuola russo-tedesca di economia matematica e la dottrina del flusso circolare." *La Scuole Economiche*. Beccatini. Turin, Utet: 387–402.

Glaisher, J. W. L., et al. 1879. *Solutions of the Cambridge Senate-House Problems and Riders for the Year 1878*. London: Macmillan.

Golinski, J. 1998. *Making Natural Knowledge: Constructivism and the History of Science*. New York: Cambridge University Press.

Golland, L. A. 1996. "Formalism in Economics." *Journal of the History of Economic Thought* 18(1): 1–12.

Golubitsky, M., and V. Guillemin 1973. *Stable Mappings and Their Singularities*. New York: Springer-Verlag.

Goodstein, J. R. 1984. "The Rise and Fall of Vito Volterra's World." *Journal of the History of Ideas* 45(4): 607–17.

Goodwin, C. D. W., ed. 1981. *Energy Policy in Perspective*. Washington, DC: Brookings Institution.

Grandmont, J.-M. 1984. "Gerard Debreu, Prix Nobel d'Economie 1983." *Societe d'etudes et de documentation économiques, industrielles et sociales* 38(Mars): 1–2.

Grattan-Guinness, I. 1990. "Does History of Science Treat of the History of Science? The Case of Mathematics." *History of Science* 28: 149–73.

——. 1994a. "From Virtual Velocities to Economic Action: The Very Slow Arrivals of Linear

Programming and Locational Analysis." In *Natural Images in Economic Thought: "Markets Read in Tooth and Claw,"* edited by P. Mirowski. New York: Cambridge University Press.

———. 1994b. "A New Type of Question": On the Prehistory of Linear and Non-linear Programming, 1770–1940." In *The History of Modern Mathematics.* Volume 3: *Images, Ideas, and Communities,* edited by E. Knobloch and D. Rowe. Boston: Academic Press. 43–89.

———. 1998. *The Norton History of the Mathematical Sciences.* New York: W. W. Norton.

———. 2000. "A Sideways Look at Hilbert's Twenty-three Problems of 1900." *Notices of the American Mathematical Society* 47(7): 752–57.

Grieve, J. 1996. "Edward Carpenter." http://www.newtel.org.uk/volunteers/jgrieve/carpent.htm.

Groenewegen, P. 1995. *A Soaring Eagle: Alfred Marshall 1842–1924.* Aldershot, UK: Edward Elgar.

Guedj, Denis. 1985. "Nicholas Bourbaki, Collective Mathematician: An Interview with Claude Chevalley," trans. Jeremy Gray. *Mathematical Intelligencer* 7(2): 18–22.

Guillemin, V., and A. Pollack 1974. *Differential Topology.* Englewood Cliffs, NJ: Prentice-Hall.

Gutting, G. 1989. *Michel Foucault's Archaeology of Scientific Reason.* New York: Cambridge University Press.

Hall, P. A., ed. 1989. *The Political Power of Economic Ideas: Keynesianism across Nations.* Princeton: Princeton University Press.

Hamermesh, D. S. 1994. "Facts and Myths about Refereeing." *Journal of Economic Perspectives* 8(1): 153–63.

Hands, D. W., and P. Mirowski. 1998. "Harold Hotelling and the Neoclassical Dream." In *Economics and Methodology: Crossing Boundaries,* edited by R. Backhouse et al. New York: Macmillan. 322–97.

Hanna, G. 1996. "The Ongoing Value of Proof." *Psychology of Mathematics Education* 20, Valencia, Spain.

Haraway, D. 1997. *Modest Witness—Second Millennium.* New York: Routledge.

Hardy, G. H. 1944 [1908]. *A Course in Pure Mathematics.* Cambridge: Cambridge University Press.

Harris, W. T. 1898. *Psychologic Foundations of Education: An Attempt to Show the Genesis of the Higher Faculties of the Mind.* New York: D. Appleton.

Henderson, J. 1996. *Early Mathematical economics: William Whewell and the British Case.* New York: Rowman and Littlefield.

Henderson, J. M., and R. E. Quandt. 1958. *Microeconomic Theory: A Mathematical Approach.* New York: McGraw-Hill.

Henderson, L. J. 1935. *Pareto's General Sociology, A Physiologist's Interpretation.* Cambridge, MA: Harvard University Press.

Heyl, B. S. 1968. "The Harvard 'Pareto Circle.' " *Journal of the History of the Behavioral Sciences* 4: 316–34.

Hicks, J. R. 1939. *Value and Capital*. Oxford: Oxford University Press.

Hilbert, D. 1918. "Axiomatisches Denken." *Mathematische Annalen* 78: 405–15.

——. 1992. *Natur und Mathematisches Erkennen: Vorlesungen, gehalten 1919–1920 in Göttingen. Nach der Ausarbeitung von Paul Bernays*. Basel: Birkhäuser.

Hildenbrand, W. 1983a. "An Axiomatic Analysis of Economic Equilibrium: On the Award of the Nobel Prize to Gerard Debreu." *Neue Zurcher Zeitung*.

Hildenbrand, W. 1983b. "Introduction." In *Mathematical Economics: Twenty Papers of Gerard Debreu* by G. Debreu. New York: Cambridge University Press. 1–29.

Hoffman, P. 1998. *The Man Who Loved Only Numbers*. New York: Hyperion.

Hookway, C. 1990. *Scepticism*. London: Routledge.

Horwich, G., and J. Pomery. 1987. "Lloyd Appleton Metzler." In *The New Palgrave: A Dictionary of Economics,* edited by J. Eatwell, M. Milgate, and N. Peter. London: Macmillan. 3: 458–61.

Howson, G. 1982. *A History of Mathematics Education in England*. Cambridge: Cambridge University Press.

Hume, D. 1949. *Treatise on Human Nature*. Volume 1. London: J. M. Dent and Sons.

Hutchison, T. W. 1938. *The Significance and Basic Postulates of Economic Theory*. London: Macmillan.

——. 1977. *Knowledge and Ignorance in Economics*. Oxford: Basil Blackwell.

Ingrao, B., and G. Israel. 1985a. "General Economic Equilibrium Theory: A History of Ineffectual Paradigm Shifts (Part 1)." *Fundamenta Scientiae* 6(1): 1–45.

——. 1985b. "General Economic Equilibrium Theory: A History of Ineffectual Paradigm Shifts (Part 2)." *Fundamenta Scientiae* 6(2): 89–125.

——. 1990. *The Invisible Hand: Economic Theory in the History of Science*. Cambridge, MA: MIT Press.

Israel, G. 1977. *Un aspetto ideologico della matematica contemporanea: il 'bourbakismo.'* Matematica e Fisica: Struttura e Ideologia, Instituto di Fisica dell'universita di Lecce, Italia, De Donato, SpA.

——. 1978. "La matematica asiomatica ed il 'bourbakismo.'" In *I Fundamenti Della matematica Dall'800 Ad Oggi*, edited by E. Casari, F. Marchetti, and G. Israel. Firenze: Guaraldi Editore SpA: 49–67.

——. 1981. "Rigor and Axiomatics in Modern Mathematics." *Fundamenta Scientiae* 2: 205–19.

——. 1982a. *Le equazioni di Volterra e Lotka: una questione di priorita*. La Storia delle matematiche in Italia, Cagliari, Italia.

——. 1982b. "Volterra Archive at the Accademia Nazionale Dei Lincei." *Historia Mathematica* 9(2): 229–38.

——. 1988. "On the Contribution of Volterra and Lotka to the Development of Modern Biomathematics." *History and Philosophy of the Life Sciences* 10: 37–49.

——. 1991a. "Volterra's 'Analytical Mechanics' of Biological Associations, First Part." *Archives Internationales D'Histoire Des Sciences* 41(126): 57–104.

—. 1991b. "Volterra's 'Analytical Mechanics' of Biological Associations, Second Part." *Archives Internationales D'Histoire Des Sciences* 41(127): 57–104.

Israel, G., and L. Nurzia. 1989. "Fundamental Trends and Conflicts in Italian Mathematics Between the Two World Wars." *Archives Internationales D'Histoire Des Sciences* 39(122): 111–43.

Jaffe, A., and F. Quinn. 1993. "Theoretical Mathematics: Toward a Cultural Synthesis of Mathematics and Theoretical Physics." *Bulletin of the American Mathematical Society (N.S.)* 29(1): 1–13.

Johanson, L. 1959. "Substitution Versus Fixed Coefficient Production Coefficients in the Theory of Economic Growth: A Synthesis." *Econometrica* 27(2): 157–76.

Kadish, A., and K. Tribe, eds. 1993. *The Market for Political Economy.* London: Routledge.

Kadvany, John. 2001. *Imre Lakatos and the Guises of Reason.* Durham: Duke University Press.

Kaluza, R. 1996. *Through a Reporter's Eyes: The Life of Stefan Banach.* Basle: Birkhäuser.

Katz, B., ed. 1989. *Nobel Laureates in Economic Sciences: A Biographical Dictionary.* New York: Garland.

Keynes, J. M. 1936. *The General Theory of Employment, Interest, and Money.* New York: Harcourt Brace.

Keyssar, A. 1986. *Out of Work : The First Century of Unemployment in Massachusetts.* Cambridge: Cambridge University Press.

Kingsland, S. 1985. *Modeling Nature.* Chicago: University of Chicago Press.

Kitcher, P., and W. Aspray. 1988. "An Opinionated Introduction." In *History and Philosophy of Modern Mathematics* by W. Aspray and P. Kitcher. Minneapolis: University of Minnesota Press. XI: 3–57.

Klamer, A. 1981. "New Classical Economics: A Methodological Examination of Rational Expectations Economics." Duke University Economics Department.

—. 1983. *Conversations with Economists.* Totowa, NJ: Roman and Allanheld.

Kleene, S. 1976. "The Work of Kurt Gödel." *Journal of Symbolic Logic* 41(4): 761–78.

—. 1988. "The Work of Kurt Gödel." In *Gödel's Theorem in Focus,* edited by S. Shanker. Beckenham, UK: Croom Helm. 48–73.

Klein, F. 1894. *The Evanston Colloquium: Lectures on Mathematics Delivered From Aug. 28 to Sept.9, 1893 before Members of the Congress of Mathematics Held in Connection with the World's Fair in Chicago at Northwestern University, Evanston, Ill. Reported by Alexander Ziwet.* New York: Macmillan.

Klein, J. L. 1998. *Statistical Visions in Time : A History of Time Series Analysis, 1662–1938.* Cambridge: Cambridge University Press.

Klein, P. A., ed. 1994. *The Role of Economic Theory. Recent Economic Thought.* Boston: Kluwer Academic.

Kline, M. 1972. *Mathematical Thought from Ancient to Modern Times.* New York: Oxford University Press.

Kline, M. 1980. *Mathematics: The Loss of Certainty.* Oxford: Oxford University Press.

Knorr-Cetina, K. 1991. "Epistemic Cultures: Forms of Reason in Science." *History of Political Economy* 23(1): 105–22.

Knorr-Cetina, K., et al., eds. 1981. *The Social Processes of Scientific Investigation*. Boston: Reidel.

Knorr-Cetina, K., and M. Mulkey, eds. 1983. *Science Observed: Perspectives on the Social Study of Science*. Beverly Hills, CA: Sage.

Koopmans, T. C. 1957. *Three Essays on the State of Economic Science*. New York: McGraw-Hill.

Kragh, H. 1987. *An Introduction to the Historiography of Science*. New York: Cambridge University Press.

Kruger, L., et al., eds. 1987. *The Probabilistic Revolution: Volume 1, Ideas in History*. Cambridge, MA: MIT Press.

—. 1987. *The Probabilistic Revolution: Volume 2, Ideas in the Sciences*. Cambridge, MA: MIT Press.

Kuhn, T. S. 1996 [1962]. *The Structure of Scientific Revolutions*. Chicago: University of Chicago Press.

Kurz, H., and N. Salvadori. 1993. "Von Neumann's Growth Model and the 'Classical' Tradition." *European Journal of the History of Economic Thought* 1(1): 129–60.

Lakatos, I. 1970. "Falsification and the Methodology of Scientific Research Programmes." In *Criticism and the Growth of Knowledge,* edited by I. Lakatos and A. Musgrave. London: Cambridge University Press. 91–196.

—. 1971. "History of Science and Its Rational Reconstructions." In *P.S. A. 1970 Boston Studies in the Philosophy of Science,* edited by R. C. Buck and R. S. Cohen. Dordrecht: Reidel. 8: 91–135.

Lang, S. 1995. "Mordell's Review, Siegel's Letter to Mordell, Diophantine Geometry, and 20th Century Mathematics." *Notices of the American Mathematical Society* 42(3): 339–50.

Latour, B. 1987. *Science in Action*. Cambridge, MA: Harvard University Press.

—. 1988. *The Pasteurization of France*. Cambridge, MA: Harvard University Press.

—. 1992. "Pasteur on Lactic Acid Yeast: A Partial Semiotic Analysis." *Configurations* 1(1): 129–45.

—. 1993. *We Have Never Been Modern*. Cambridge, MA: Harvard University Press.

Latsis, S., ed. 1976. *Method and Appraisal in Economics*. New York: Cambridge University Press.

Lax, P. 1989. "The Flowering of Applied Mathematics in America." In *A Century of Mathematics in America,* edited by P. Duren. Providence, RI: American Mathematical Society. II: 455–66.

Leonard, R. 1992. "Creating a Context for Game Theory." In *Toward a History of Game Theory,* edited by E. R. Weintraub. Durham, NC: Duke University Press.

—. 1994. "Reading Nash, Reading Cournot: The Creations and Stabilisation of the Nash Equilibrium." *Economic Journal* 104(424): 492–511.

—. 1995. "From Parlor Games to Social Science: von Neumann, Morgenstern, and the Creation of Game Theory 1928-1944." *Journal of Economic Literature* 33: 730–61.

—. Forthcoming. *From Red Vienna to Santa Monica: von Neumann, Morgenstern, and Social Science, 1925–1955.* Cambridge: Cambridge University Press.

Littlewood, J. E. 1953. *A Mathematician's Miscellany.* London: Methuen.

Macrae, N. 1992. *John von Neumann.* New York: Pantheon.

Mandelbrot, B. 1989. "Chaos, Bourbaki, and Poincarè." *Mathematical Intelligencer* 11(3): 10–12.

Marshall, A. 1949. *Principles of Economics.* New York: Macmillan.

Mathias, A. "The Ignorance of Bourbaki." *Mathematical Intelligencer* 14(3): 4–13.

Mazeres, R. 1998. Telephone conversation, 2 January.

McCloskey, D. N. 1986. *The Rhetoric of Economics.* Madison: University of Wisconsin Press.

—. 1988. "Thick and Thin Methodologies in the History of Economic Thought." In *The Popperian Legacy in Economics,* edited by N. B. DeMarchi. Cambridge: Cambridge University Press. 245–57.

—. 1994. *Knowledge and Persuasion in Economics.* New York: Cambridge University Press.

—. 1999. *Crossing : A Memoir.* Chicago: University of Chicago Press.

McCormmach, R. 1982. *Night Thoughts of a Classical Physicist.* New York: Avon.

Mehrtens, H. 1981. "Social History of Mathematics." In *Social History of Nineteenth Century Mathematics,* edited by H. Mehrtens, H. Bos, and I. Schneider. Boston: Birkhäuser. 257–80.

—. 1990. *Moderne—Sprache—Mathematik.* Frankfurt am Main: Suhrkamp.

Menger, K. 1973. "Austrian Marginalism and Mathematical Economics." In *Carl Menger and the Austrian School of Economics,* edited by J. R. Hicks and W. Weber. Oxford: Oxford University Press. 38–60.

Miller, W. G. 1952. "The Mathematics of Production and Consumption in a Static Economy." University of Florida Department of Mathematics.

Mirowski, P. 1989. *More Heat Than Light.* New York: Cambridge University Press.

—. 1990. "Problems in the Paternity of Econometrics: Henry Ludwell Moore." *History of Political Economy* 22(4): 587–610.

—. 1992. "What Were von Neumann and Morgenstern Trying to Accomplish?" In *Toward a History of Game Theory,* edited by E. R. Weintraub. Durham, NC: Duke University Press.

—. 1994. "Marshalling the Unruly Atoms: Understanding Edgeworth's Career." In *Edgeworth on Chance, Economic Hazard, and Statistics* by P. Mirowski. London: Rowman and Littlefield. 1–80.

—. 1999. *Machine Dreams.* Cambridge, MA: Harvard University Press.

Moore, H. L. 1929. *Synthetic Economics.* New York: Macmillan.

Mordell, L. J. 1964. "Review of Lang's Diophantine Geometry." *Bulletin of the American Mathematical Society* 70: 491–98.

Morgan, M. 1990. *The History of Econometrics.* Cambridge: Cambridge University Press.

Morgan, M. S. 1999. "Models, Stories, and the Economic World." Amsterdam: University of Amsterdam, Faculty of Economics and Econometrics.

Morgan, M. S., and M. Morrison, eds. 1999. *Models as Mediators. Ideas in Context.* New York: Cambridge University Press.

Moritz, R. E. 1914. *Memorabilia Mathematica*. Washington, DC: Mathematical Association of America.

Morrey, C. B. J. 1983. "Griffith Conrad Evans." *Biographical Memoirs: U. S. National Academy of Sciences* 54: 127–55.

Mumford, D. 1991. "A Foreword for Non-Mathematicians." In *The Unreal Life of Oscar Zariski,* edited by C. A. N. E. Parikh. San Diego: Academic Press. xv–xxvii.

Nasar, S. 1998. *A Beautiful Mind.* New York: Simon and Schuster.

Novick, P. 1988. *That Noble Dream: The "Objectivity Question" and the American Historical Profession.* Cambridge: Cambridge University Press.

Osgood, W. F. 1925. *Advanced Calculus.* New York: Macmillan.

Pareto, V., and J. I. Griffin. 1955 [1911]. "Mathematical Economics." *International Economic Papers* 5: 58–102.

Parshall, K. H., and D. E. Rowe. 1994. *The Emergence of the American Mathematical Research Community, 1876–1900: J. J. Sylvester, Felix Klein, and E. H. Moore.* Providence, R I: American Mathematical Society.

Patinkin, D. 1948a. "Price Flexibility and Full Employment." *American Economic Review* 38: 543–64.

———. 1948b. "Relative Prices, Say's Law, and the Demand for Money." *Econometrica* 16: 135–54.

———. 1949. "The Indeterminacy of Absolute Prices in Classical Theory." *Econometrica* 17: 1–27.

———. 1951. "The Invalidity of Classical Monetary Theory." *Econometrica* 19: 134–51.

———. 1965. *Money , Interest, and Prices.* New York: Harper and Row.

———. 1981. *Essays On and In the Chicago Tradition.* Durham, NC: Duke University Press.

———. 1995. "The Training of an Economist." *Banca Nazionale del Lavoro Quarterly Review* 48(195): 359–95.

Pearson, K. 1911. *The Grammar of Science.* New York: Macmillan.

Peckhaus, V. 1990. *Hilbertprogramm und Kritische Philosophie.* Göttingen: Vandenhoeck and Ruprecht.

———. 1994. "Hilbert's Axiomatic Programme and Philosophy." In *The History of Modern Mathematics, Volume III: Images, Ideas, and Communities,* edited by E. Knobloch and D. Rowe. Boston: Academic Press. 91–112.

Phipps, C. G. 1928. "Problems in Approximation by Functions of Given Continuity." Minneapolis-St. Paul: University of Minnesota Mathematics Department.

———. 1950a. "A Note on Patinkin's 'Relative Prices.' " *Econometrica* 18(1): 25–26.

———. 1950b. "A Note on Tintner's 'Homogeneous Systems.' " *Econometrica* 18(1): 63.

———. 1952a. "Friedman's 'Welfare' Effects." *Journal of Political Economy* 60(4): 332–34.

———. 1952b. "Money in the Utility Function." *Metroeconomica* 4: 44–65.

Pickering, A. 1995. *The Mangle of Practice: Time, Agency, and Science.* Chicago: University of Chicago Press.

Pinch, T. J. 1977. "What Does A Proof Do If It Does Not Prove?" In *The Social Production of*

Scientific Knowledge, edited by E. Mendelsohn, P. Weingart, and R. Whitley. Dordrecht: D. Reidel. 1: 171–215.

Podhoretz, N. 1967. *Making It.* New York: Random House.

Popper, K. 1959 [1934]. *The Logic of Scientific Discovery.* London: Hutchinson.

Porter, T. M. 1992. "Comment on Schabas." *History of Political Economy* 24(1): 234–36.

——. 1995. *Trust in Numbers: The Pursuit of Objectivity in Science and Public Life.* Princeton: Princeton University Press.

Price, G. B. 1988. "American Mathematicians in World War I." In *A Century of Mathematics in America,* edited by P. Duran. Providence, RI: American Mathematical Society. 1: 267–68.

Punzo, L. 1991. "The School of Mathematical Formalism and the Viennese Circle of Mathematical Economics." *Journal of the History of Economic Thought* 13(1): 1–18.

Reid, P. 1996. *Art and Affection: A Life of Virginia Woolf.* New York: Oxford University Press.

Reingold, N. 1991. "The Peculiarities of the Americans, or Are There National Styles in the Sciences." *Science in Context* 4(2): 347–66.

Richards, J. L. 1988. *Mathematical Visions: the Pursuit of Geometry in Victorian England.* San Diego: Academic Press.

——. 1991. "Rigor and Clarity: Foundations of Mathematics in France and England, 1800–1840." *Science in Context* 4(2).

Rider, R. E. 1989. "An Opportune Time: Griffith C. Evans and mathematics at Berkeley." In *A Century of Mathematics in America,* edited by P. Duren. Providence, RI: American Mathematical Society. 2: 283–302.

Roos, C. F. 1925. "A Mathematical Theory of Competition." *American Journal of Mathematics* 47(July): 163–75.

Rorty, R. 1980. *Philosophy and the Mirror of Nature.* Oxford: Basil Blackwell.

——. 1984. "The Historiography of Philosophy: Four Genres." In *Philosophy in History,* edited by R. Rorty, J. B. Schneewind, and Q. Skinner. Cambridge: Cambridge University Press. 49–75.

——. 1998. *Truth and Progress: Philosophical Papers.* Vol. 3. New York: Cambridge University Press.

Ross, D. 1991. *The Origins of American Social Science.* New York: Cambridge University Press.

Roth, L. 1971. "Old Cambridge Days." *American Mathematical Monthly* 78: 223–26.

Rothrock, D. A. 1919. "American Mathematicians in War Service." *American Mathematical Monthly* 26: 40–44.

Rouse, J. 1993. "What Are Cultural Studies of Scientific Knowledge?" *Configurations* 1(1): 1–22.

Rowe, D. E. 1994. "The Philosophical Views of Klein and Hilbert." In *The Intersection of History and Mathematics,* edited by S. Chikara, S. Mitsuo, and J. W. Dauben. Boston: Birkhäuser.

Rowe, D. E. 1997. "Perspective on Hilbert." *Perspectives on Science* 5(4): 533–70.

Russett, C. E. 1966. *The Concept of Equilibrium in American Social Thought.* New Haven: Yale University Press.

Samuels, W. J. 1998. "Journal Editing in the History of Economic Thought." *History of Economics Review* 27(Winter): 3–5.

Samuelson, P. A. 1941. "The Stability of Equilibrium: Comparative Statics and Dynamics." *Econometrica* 9: 97–120.

—. 1945. "The Stability of Multiple Markets: The Hicks Conditions." *Econometrica* 13(4): 277–92.

—. 1947. *Foundations of Economic Analysis.* Cambridge, MA: Harvard University Press.

—. 1983. *Foundation of Economic Analysis (Enlarged Edition).* Cambridge, MA: Harvard University Press.

—. 1987. "Out of the Closet: A Program for the Whig History of Economic Science." *History of Economics Society Bulletin* 9(1): 51–60.

—. 1988. "Keeping Whig History Honest." *History of Economics Society Bulletin* 10(2): 161–67.

Sarton, G. 1936. *The Study of the History of Science.* Cambridge, MA: Harvard University Press.

Schabas, M. 1992. "Breaking Away: History of Economics as History of Science." *History of Political Economy* 24(1): 187–203.

Shackle, G. L. S. 1967. *The Years of High Theory.* Cambridge: Cambridge University Press.

Seidel, E. 1990. "Gravitational Radiation from Even-Parity Perturbations of Stellar Collapse: Mathematical Formalism and Numerical Methods." *Physical Review* 42(1884).

Sent, E.-M. 1996). "What an Economist Can Teach Nancy Cartwright." South Bend, IN: Notre Dame University, Department of Economics.

—. 1998. *The Evolving Rationality of Rational Expectations: An Assessment of Thomas Sargent's Achievements.* Cambridge: Cambridge University Press.

Shanker, S., ed. 1988. *Gödel's Theorem in Focus.* Beckenham, UK: Croom Helm.

Shapin, S. 1994. *A Social History of Truth.* Chicago: University of Chicago Press.

—. 1996. *The Scientific Revolution.* Chicago: University of Chicago Press.

Shapin, S., and S. Schaffer. 1985. *Leviathan and the Air-Pump.* Princeton: Princeton University Press.

Shepherd, G. B., ed. 1995. *Rejected: Leading Economists Ponder the Publication Process.* Sun Lakes, AZ: Thomas Horton and Daughters.

Sigurdsson, S. 1992. "17,000 Reprints Later: Description and analysis of the Vito Volterra Reprint Collection." 22(2): 391–97.

Simon, H. 1959. "Review of 'Elements of Mathematical Biology.'" *Econometrica* 27(3): 493–95.

Simon, H. A. 1987. "Evans, Griffith Conrad." In *The New Palgrave: A Dictionary of Economics,* edited by J. Eatwell, M. Milgate, and P. Newman. New York: Stockton Press. 2: 198–99.

Skidelsky, R. 1986. *John Maynard Keynes, Hopes Betrayed 1883–1920.* New York: Viking.

Smale, S. 1972. "Personal Perspectives on Mathematics and Mechanics." In *Statistical Mechanics: New Concepts, New Problems, New Applications. Proceedings of the Sixth IUPAP Conference on Statistical Mechanics,* edited by Rice, Light, and Freed. Chicago: University of Chicago Press.

—. 1976. "Dynamics in General Equilibrium." *American Economic Review* 66(2): 284–94.

—. 1977. *Some Dynamical Questions in Mathematical Economics.* Paris: Centre National de la Recherche Scientifique.

—. 1980. *The Mathematics of Time.* New York: Springer-Verlag.

Smith, B. H. 1988. *Contingencies of Value.* Cambridge, MA: Harvard University Press.

—. 1997. *Belief and Resistance: Dynamics of Contemporary Intellectual Controversy.* Cambridge, MA: Harvard University Press.

Stigler, G. J. 1969 [1949]. *Five Lectures on Economic Problems.* Freeport, NY: Books for Libraries Press.

Stigler, S. 1986. *The History of Statistics.* Cambridge, MA: Harvard University Press.

Stoltz, M. P. 1951. "Review of 'A Note on Patinkin's Relative Prices.'" *Mathematical Reviews* 11: 11530g.

Stone, M. H. 1989. "Reminiscences of Mathematics at Chicago." In *A Century of Mathematics in America, Part II,* edited by P. Duren. Providence: American Mathematical Society.

Struik, D. J. 1942. "The Sociology of Mathematics." *Science and Society* 6(1): 58–70.

Stump, D. J. 1997. "Reconstructing the Unity of Mathematics circa 1900." *Perspectives on Science* 5(3): 383–417.

Sturges, R. P. 1975. *Economists' Papers: A Guide to Archive and Other Manuscript Sources for the History of British and Irish Economics Thought.* Durham, NC: Duke University Press.

Szenberg, M., ed. 1992. *Eminent Economists: Their Life Philosophies.* New York: Cambridge University Press.

Tapon, F. 1973. "A Contemporary Example of the Transition to Maturity in the Social Sciences: The Peculiar State of Economics in France." Durham, NC: Duke University, Department of Economics.

Thom, R. 1975 [1972]. *Structural Stability and Morphogenesis.* Reading, MA: W. A. Benjamin.

Thompson, J. J. 1936. *Recollections and Reflections.* London: G. Bell and Sons.

Tintner, G. 1948. "Homogeneous Systems in Mathematical Economics." *Econometrica* 16: 273–94.

Toepell, M. M. 1986. *Über die Entstehung von David Hilberts 'Grundlagen der Geometrie.'* Göttingen: Vandenhoeck and Ruprecht.

Varian, H. 1984. "Gerard Debreu's Contribution to Economics." *Scandinavian Journal of Economics* 86(1): 4–14.

—. 1995. Letter to E. Roy Weintraub.

Vinner, S., and D. Tall. 1982. "Existence Statements and Constructions in Mathematics and Some Consequences to Mathematics Teaching." *American Mathematical Monthly* 89(10): 752–56.

Volkmann, P. 1900. *Einführung in das Studium der theoretischen Physik, insbesondere das der analytischen Mechanik mit einer Einleitung in die Theorie der Physikalischen Erkenntniss.* Leipzig: Teubner.

Volterra, E. 1976. "Volterra, Vito." In *Dictionary of Scientific Biography*, edited by C. C. Gillespie. New York: Charles Scribner's Sons. 14: 85–88.

Volterra, V. 1906a. "L'Economia matematica ed il Nuovo manuale del Prof. Pareto." *Giornale degli economisti* 32(serie II): 296–301.

———. 1906b. "Les mathématiques dans les sciences Biologiques et Sociales." *La Revue du Mois* 1(10 janvier): 1–20.

———. 1915. "Henri Poincarè: A Lecture Delivered at the Inauguration of the Rice Institute." *Rice Institute Pamphlet* 1(2): 133–62.

———. 1957. *Opere Matematiche: Memorie e Note*. Roma: Accademia Nazionale Dei Lincei.

———. 1959. *Theory of Functionals and of Integral and Integro-differential Equations*. New York: Dover.

von Neumann, J. 1928. "Zur Theorie der Gesellschaftsspiele." *Mathematische Annalen* 100: 295–320.

———. 1936. "Über ein ökonomisches Gleichungssystem und eine Verallgemeinerung des Brouwerschen Fixpunksätzes." In *Ergebnisse eines Mathematischen Kolloquiums 1935–36*, edited by K. Menger. Leipzig: Franz Deuticke. 8: 73–83.

———. 1947. "The Mathematician." In *The Works of the Mind*, edited by R. B. Heywood. Chicago: University of Chicago Press. 180–96.

von Neumann, J., and O. Morgenstern. 1944. *The Theory of Games and Economic Behavior*. Princeton: Princeton University Press.

Walras, L. 1954. *Elements of Pure Economics*. London: Allen and Unwin.

Walton, K. 1990. "Is Nicholas Bourbaki Alive?" *Mathematics Teacher* (November): 666–68.

Warwick, A. 1998. "Exercising the Student Body: Mathematics and Athleticism in Victorian Cambridge." In *Science Incarnate: Historical Embodiments of Natural Knowledge*, edited by C. Lawrence and S. Shapin. Chicago: University of Chicago Press. 288–326.

Weil, A. 1978. *History of Mathematics: Why and How*. Helsinki: International Congress of Mathematicians, Academia Scientiarum Fennica.

Weintraub, E. R. 1969. "Stability of Stochastic General Equilibrium Systems." *Applied Mathematics*. Philadelphia: University of Pennsylvania. xii + 116.

———. 1975. "Uncertainty and the Keynesian Revolution." *History of Political Economy* 7(4): 530–48.

———. 1979. *Microfoundations: The Compatibility of Microeconomics and Macroeconomics*. New York: Cambridge University Press.

———. 1985. *General Equilibrium Analysis: Studies in Appraisal*. New York: Cambridge University Press.

———. 1988. "The Neo-Walrasian Program is Empirically Progressive." In *The Popperian Legacy in Economics*, edited by N. DeMarchi. Cambridge: Cambridge University Press. 213–27.

———. 1990. "Comment on Heilbroner." In *Economics as Discourse*, edited by W. Samuels. Boston: Kluwer. 117–28.

——. 1991a. "Allais, Stability, and Liapunov Theory." *History of Political Economy* 23(3): 383–96.

——. 1991b. "From Dynamics to Stability." In *Appraising Economic Theories: Studies in the Methodology of Scientific Research Programs,* edited by M. Blaug and N. DeMarchi. Cheltenham, UK: Edward Elgar. 273–91.

——. 1991c. *Stabilizing Dynamics.* New York: Cambridge University Press.

——. 1991d. "Surveying Dynamics." *Journal of Post Keynesian Economics* 13(4): 525–44.

——. 1992a. "Commentary: Historical Case Studies are Made, Not Given." In *Post-Popperian Methodology in Economics,* edited by N. DeMarchi. Boston: Kluwer Academic. 355–74.

——. 1992b. "Introduction." In *Toward a History of Game Theory,* edited by E. R. Weintraub. Durham, NC: Duke University Press. 3–12.

——. 1992c. "Roger Backhouse's Straw Herring." *Methodus* 4(2): 53–57.

——. 1992d. "Thicker is Better." *Journal of the History of Economic Thought* 14(2): 271–76.

——. 1998. "Axiomatisches Mißverständnis." *Economic Journal* 108(November): 1837–47.

——. 1998). "From Rigor to Axiomatics: The Marginalization of Griffith C. Evans." In *On the Transformation of American Economics, From Prewar Pluralism to Postwar Neoclassicism,* edited by M. Morgan and M. Rutherford. Durham, NC: Duke University Press.

Weintraub, E. R., ed. 1992. *Toward a History of Game Theory.* Durham, NC: Duke University Press.

Weintraub, E. R., and P. Mirowski. 1994. "The Pure and the Applied: Bourbakism Comes to Mathematical Economics." *Science in Context* 7(2): 245–72.

Weintraub, S. 1958. *An Approach to the Theory of Income Distribution.* Philadelphia: Chilton.

——. 1989. "A Jevonian Seditionist: A Mutiny to Enhance the Economic Bounty?" In *Recollections of Eminent Economists,* Volume 1, edited by J. A. Kregel. New York: New York University Press. 1: xix + 234.

Westfall, R. S. 1980. *Never at Rest.* New York: Cambridge University Press.

Weyl, H. 1944. "David Hilbert and his Mathematical Work." *Bulletin of the American Mathematical Society* 50: 612–54.

Whewell, W. 1845. *Of a Liberal Education in General; and with Particular Reference to the Leading Studies of the University of Cambridge.* London: J. W. Parker.

Whitaker, J. K., ed. 1975. *The Early Economic Writings of Alfred Marshall: 1867–1890,* Volume 1. London: Macmillan.

——. 1996. *The Correspondence of Alfred Marshall, Economist.* Cambridge: Cambridge University Press.

White, H. 1973. *Metahistory.* Baltimore: Johns Hopkins University Press.

——. 1990. *Tropics of Discourse.* Baltimore: Johns Hopkins University Press.

Whitehead, A. N. 1917. "Presidential Address: Mathematics and Physics Section." *Report of the Eighty-sixth Meeting of the BAAS held at Newcastle in August 1916.* London: John Murray.

Whittaker, E. T. 1959. "Biography of Vito Volterra, 1860–1940." In *Theory of Functionals and of Integral and Integro-differential Equations* by V. Volterra. New York: Dover. 5–39.

Winstanley, D. A. 1947. *Later Victorian Cambridge*. Cambridge: Cambridge University Press.

Wise, M. N., ed. 1995. *The Values of Precision*. Princeton: Princeton University Press.

Wittgenstein, L. 1958. *Philosophical Investigations*. Oxford: Basil Blackwell.

———. 1984. *Culture and Value*. Chicago: University of Chicago Press.

Wittman, W. 1967. "Die extremale Wirtschaft, Robert Remak—ein Vorläufer der Aktivität-sanalyse." *Jahrbücher für Nationalökonomie und Statistik* 180: 397–409.

Woo, H. K. H. 1986. *What's Wrong With Formalization in Economics? An Epistemological Critique*. Newark, CA: Victoria Press.

Woodcock, A., and M. Davis 1978. *Catastrophe Theory*. New York: E. P. Dutton.

Woolgar, S. 1988. *Science: The Very Idea*. London: Tavistock.

Woyczynski, W. 1996. "Appendix 1: Mathematics in Stefan Banach's Time." In *Through a Reporter's Eye: The Life of Stefan Banach*, edited by R. Kaluza. Basel: Birkhäuser.

Zeeman, E. C. 1977. *Catastrophe Theory: Selected Papers, 1972–1977*. Reading, MA: Addison-Wesley.

Zuckerman, H., and R. K. Merton. 1971. "Patterns of Evaluation in Science: Institutionalization, Structure and Functions of the Referee System." *Minerva* 9: 66–100.

Index

Library of Congress Cataloging-in-Publication Data
Weintraub, E. Roy.
How economics became a mathematical science / E. Roy Weintraub.
p. cm. — (Science and cultural theory)
Includes bibliographical references and index.
ISBN 0-8223-2856-9 (cloth : alk. paper) — ISBN 0-8223-2871-2 (pbk. : alk. paper)
1. Economics, Mathematical—History. I. Title. II. Series.
HB135 .w437 2002   330'.01'51—dc21   2001007280